William M. Gabb

On the Indian Tribes and Languages of Costa Rica

William M. Gabb

On the Indian Tribes and Languages of Costa Rica

ISBN/EAN: 9783743395381

Manufactured in Europe, USA, Canada, Australia, Japa

Cover: Foto ©berggeist007 / pixelio.de

Manufactured and distributed by brebook publishing software
(www.brebook.com)

William M. Gabb

On the Indian Tribes and Languages of Costa Rica

ON THE
INDIAN
TRIBES AND LANGUAGES
OF
COSTA RICA.

BY WM. M. GABB.

(Read before the American Philosophical Society, Aug. 20, 1875.)

PHILADELPHIA:
McCALLA & STAVELY, PRINTERS, NOS. 237-9 DOCK ST.
1875.

ON THE
INDIAN TRIBES AND LANGUAGES
OF COSTA RICA.

BY WM. M. GABB.

(*Read before the American Philosophical Society, August 20, 1875.*)

CHAPTER 1.

GENERAL ETHNOLOGICAL NOTES.

The Indians of Costa Rica, with the hardly probable exception of the Guatusos, all belong to one closely allied family. I only make this possible exception in deference to the almost absolute ignorance which yet exists in regard to this isolated tribe.

Before entering on the consideration of the better known peoples of the southern part of the Republic, it may be as well to make a brief summary of what is known of the Guatusos up to the present time. They occupy a part of the broad plains north and east of the high volcanic chain of North-Western Costa Rica, and south of the great lake of Nicaragua, especially about the head waters of the Rio Frio. I have fortunately fallen in with various persons who have entered their country, and who have had an opportunity of seeing the people and their mode of life. The stories of some are so evidently exaggerated that I shall suppress them; but by carefully sifting the evidence and giving a due preponderance to the testimony of those whom I consider most reliable, I have arrived at the following results.

Thomas Belt, the author of "The Naturalist in Nicaragua," says he has seen of them, five children and one large boy, "and they all had the common Indian features and hair; though it struck me that they appeared rather more intelligent than the generality of Indians." He also says that "one little child that Dr. Seeman and I saw in San Carlos in

1870, had a few brownish hairs among the great mass of black ones; but this character may be found among many of the indigenes, and may result from a very slight admixture of foreign blood." All the persons with whom I have conversed assert that the name Guatuso, as applied to the tribe, is given on account of a reddish or brown tint of their hair, resembling the little animal of that name (the *Agouti*). This is also denied by Mr. Belt, who says that the names of animals are often applied to Indian tribes by their neighbors, to distinguish them. Allowing full weight to this opinion, supported by analogy as it is in North America, (*e. g. Snakes*,) I do not think it fully warranted in this case.

Of half a dozen persons with whom I have conversed; people who have been on the upper Rio Frio, all, with one exception, distinctly assert that they have seen people of light color and with comparatively light hair among them. One person went so far as to assert, that in a fracas in which he nearly lost his life, his most valiant and dangerous opponent was a young woman, a mere girl, "as white as an Englishwoman," (tan rubia como una Inglesa). Another, who had a more peaceful opportunity of seeing a party of two or three women, himself unseen, used the same words in describing one of them. I believe, however, that these were exaggerations. Still another person told me that they were of all shades "from a rather light Indian color, to nearly white, the same as ourselves" (referring to the varying shades in the mixed blood of the Costa Rican peasantry). However, in an interesting conversation with Don Tomas Guardia, President of Costa Rica, I learned that when, some years ago, he headed a party passing through their country for military purposes, they encountered one or more bodies of these people and had some skirmishes with them. He says they are ordinarily of the color of other Indians, although rare exceptions exist, of individuals markedly lighter than the others, and really possessing a comparatively white skin and brownish or reddish hair. This is in keeping with the statements made to me by others whom I consider reliable, and must, I think, in deference to the authors be taken as final.

The origin of light complexions among an isolated tribe of Indians has, of course, been the source of much speculation, but General Guardia, and Don Rafael Acosta, an intelligent gentleman of San Ramon, not far from the borders of the Guatuso country, both suggested to me, independently, the same theory. They claim that when, a couple of centuries ago, the town of Esparza was sacked by the English freebooters, many of the inhabitants took refuge in the mountains, and were never afterwards heard of. These refugees were many of them pure whites, men and women. Now from Esparza, it is only about three or four days' journey to the borders of the Guatuso country, and it does not seem improbable that some of these poor wretches may have found their way there. If this is really the case, the admixture of blood, and consequent lightening of color is satisfactorily accounted for.

In consequence of almost uniform bad treatment, robbery and massacre included, to which these people have been subjected by the rubber hunters, who enter their country from Nicaragua, and their not possessing fire-arms to repel the aggressors, they have become so timid that they fly on the first approach of strangers. The few who have been captured are either young children, or persons taken by surprise. I have been unable to learn of any in Costa Rica, although a boy, now dead, lived for a while in Alajuela. A few are said to have been taken to San Juan del Norte, (Greytown,) and to Grenada, Nicaragua. The Alajuela boy, although he learned the meaning of some Spanish words, so as to know what was meant, when spoken to, was represented as sullen. When asked the names in his language of things that he was familiar with, like plantain, banana, &c., he always remained silent, and neither coaxing nor threats could extort a word.

The people are invariably represented as of short stature, broad, and of enormous strength. They live in neighborhoods; they cannot be called villages, the houses being scattered over an extensive area and at distances of from one to several hundred yards apart. The houses are low, consisting of a roof, pitching both ways from a ridge pole, and resting on very short but very thick posts. This is thatched with palm leaf and is entirely open at the ends and sides, under the eaves. Their tools are stone axes set in wooden handles, good steel machetes (all agree that they have seen these, but where do they get them?) and planting sticks similar to those used by the Bri-bris. With these tools they cultivate great quantities of plantains, bananas, yuca, coco, (*Colocasia esculentum,*) besides possessing large plantations of the *pehi balla* palm and of cacao. Of the furniture in their houses, I was told of cord hammocks and net bags, similar to those of Bri-bri, and of blocks of light wood for seats. They seem to sleep on the ground floor of their houses, simply spreading down a layer of plantain leaves. Their bows and arrows are described as similar to what I have seen elsewhere, except that the arrows are not supplied with any harder points than those furnished by the pehi balla wood. The dress is described as identical with the old styles in Talamanca; mastate breech cloths for the men, and the same material, in the shape of short petticoats for the women.

The country of the Rio Frio is said to consist of broad fertile plains, unsurpassed in beauty and fertility by any lands in the Republic. The Rio Frio itself is large and is navigated by the large canoes of the *huleros*, or rubber hunters, to a point within three days' walk of Las Cruces on the Pacific side. But the poor inoffensive people who inhabit this region are now so intimidated by the "Christians" who have visited them, that they can only be approached by a foreigner by stealth. If they can escape they do so, but if driven to bay, or think they can overpower the strangers, they greet them with a flight of arrows. They are especially afraid of firearms, and a pistol shot is sufficient to depopulate a settlement.

I believe the above short statement contains the most reliable information ever yet accumulated with reference to the Guatusos. I have carefully rejected many wonderful stories told me by persons claiming to tell what they saw, and have only availed myself of the accounts of those who seemed to exaggerate least, or whose position forbade me to doubt their assertions.

The tribes of Southern and South-eastern Costa Rica are better known. The Terrabas, living on the Pacific slope, and their neighbors, the Borucas or, as they call themselves, Bruncas, live under complete subjection to the laws of Costa Rica, and the rule of a missionary priest. They may be strictly called civilized. But those on the Atlantic slope have had a powerful ally in the forces of nature, in resisting the civilizing efforts of the Spanish invaders. The heavy rains of the Atlantic seaboard produce a luxuriance of vegetation that may well nigh be called unconquerable. Broad swamps, dank and reeking with malaria threaten the European with bilious fever, fatal to energy if not to life. Three centuries ago Columbus sailed along the coast from the Bahia del Almirante, and in his usual florid style called this the Rich Coast, and yet it has never yielded to the conqueror or paid him tribute. Two centuries ago a little colony was planted far back in the mountains and one or two outlying missionary posts were scattered among the then powerful tribes. But a just retribution fell on San José de Cabecar. The hardy mountaineers did not submit to the oppressors' yoke like the gentle and hapless victims of Cuba and Santo Domingo. Even now the traditions are well preserved among them, and I have listened to more than one recital of outrages which I dare not believe to be exaggerated. Father Las Casas tells of even worse oppressions. In 1709 the people rose and massacred all who fell into their power. A pitiful remnant escaped from the colony, to wander for weeks in the woods and finally a handful reached Cartago. The Viceroy of Guatemala, in retaliation sent forces by way of the forest trails from Cartago and others across the mountains by way of Terraba. They surrounded, killed, and captured all the Indians they could, and carried their prisoners to Cartago. Some of these were divided among the settlers as servants, and have left a strong tinge on the cheeks of many a would-be high-toned Costa Rican. The remainder were settled in the villages of Tucuriqui and Orosi, where, though partly civilized, they still retain their original language, badly corrupted with Spanish. Since this disastrous ending to the colony, both parties have kept up a wholesome dread of each other and no further efforts have ever been made to found a colony on the Atlantic side of the country. At the same time, the Indians not only dread, but hate the Spaniards and even a trace of Spanish blood, or fluency in the language on the part of a dark-skinned or dark-haired person is a warrant for suspicion. It is not a hatred of the white race. Englishmen, Americans, and Germans are invariably respected and treated well, by the same people who are either insolent to the Spaniard or treat him at best with restraint.

On the Atlantic slope, there are three tribes intimately allied socially, politically, and religiously, but differing markedly in language. The Cabecars occupy the country from the frontiers of civilization to the western side of the Coen branch of the Tiliri or Sicsola River. Adjoining them, the Bri-bris occupy the east side of the Coen, all the regions of the Lari, Uren, and Zhorquin and the valley lying around the mouths of these streams. The Tiribris, now reduced to barely a hundred souls, live in two villages on the Tilorio or Changinola River. It is said that on the head waters of the Changina, a large fork of this latter stream, there are yet a few individuals of the Changina tribe, but the other Indians report them as implacably hostile and their very existence is only known by vague reports of their savage neighbors. The Shelaba tribe, formerly living on the lower part of the same river is now entirely extinct. A few half-breeds are all who perpetuate the blood, and their language is utterly lost. Still further down the coast, beyond the Costa Rican boundaries is another allied tribe, partly civilized, in so far as that they trade and work a little and drink a great deal of bad rum, spending most of their earnings on that bane of the race. They are called by foreigners Valientes. Crossing over to the Pacific slope, the Terrabas are tribally identical with the Tiribis. The tradition still exists in a vague form, that they are emigrants from the Atlantic side; but when or why the emigration took place, is forgotten. The home of the tribe is in a very narrow, rough cañon, traversed by a river that might better be called a torrent, a country strongly contrasted with the fertile plains and broad savannas of Terraba, and it is not improbable that under the press of a crowded population several migrations took place. They still tell how, twenty or thirty years ago, a priest came over from Terraba, baptized all who would submit to the rite, and by glowing stories of the abundance of meat and other inducements that he shrewdly imagined would tempt them, carried off over a dozen of their best men, who never returned. A glance at the vocabulary will show how little separated are these two branches of the tribe in language. The Borucas or Bruncas, who occupy a little village, not far from the headquarters of the Terrabas, are apparently the older occupants of the soil; perhaps crowded into a corner by the invaders.

Other tribal names are mentioned by various authors, such as Biceitas, &c. The name Biceita is not known in the country, and, although used to the present day outside of the Indian country, is unknown to them, or at best, is supposed to be a Spanish word. The district of that name is probably the western part of Bri-bri, the most eastern point to which the slave-hunting expeditions from San José Cabecar penetrated. The Blancos are properly the Bri-bri tribe, but this word is rather loosely used, and is often applied alike to the Cabecars and Tiribis.

But little can be gathered of the history of these people. What happened in the times of their grandfathers is already ancient history and partly forgotten. All recollection of the first arrival of the Spaniards is now lost. They have no traditions of the use of stone implements before the introduction of metal. When asked what they did for axes before the traders came among them, I could get no more satisfactory answer than that they went to Cartago to buy them. I have been told a vague story, however, that long ago there were two bands living in the country now occupied by the Bri-bris. Those living in the valley, around the junction of the branches of the Tiliri were more powerful than the mountaineers, and forced the latter to pay tribute when they descended to hunt, or cut the material for their bark-cloth clothing. But gradually the lowlanders died out: the highlanders, becoming the more powerful, rebelled against these impositions, and eventually emigrated in such numbers to the country of the former, that the distinction became lost by an amalgamation of the two parties. Even now the Bri-bris, who occupy the lowlands and most of the hill regions of the Sicsola, look down on their neighbors the Cabecars and treat them as inferiors. The Cabecars, on the other hand, tacitly acknowledge even a social supremacy, and in a mixed party submit to assume the more menial occupations, like bringing water and wood; and are always obliged to wait until the last when food or drink is being served. Few of the Bri-bris speak the Cabecar language, but there are few of the Cabecars who do not speak Bri-bri, and they usually use it in the presence of strangers. The Cabecars have no chief of their own, but are entirely under the rule of the Bri-bri chief, and have been, from time immemorial. Their subjugation is, in short, complete. At the same time they have the honor of religious supremacy, in so far as that the high priest, the "*Usekara*," whose office will be explained further on, belongs to their tribe. The ordinary priests, the "*Tsugurs*," who, like the "*Usekara*," are hereditary, come from a group of families on the Coen River, but belong to the Bri-bri tribe.

About the beginning of this century there was a bitter war between the Bri-bris and the Tiribis. The youngest members of the war parties are now mostly dead, and the few remaining survivors are very old men. The last of the warriors proper, mature men at that time, died about 1860, at an extremely advanced age. I have heard the traditions from both sides the question, and of course each party throws all the blame on the other. The Bri-bri story is that some people, a whole family, living on the extreme eastern portion of the Uren district, were found murdered, and no clue discovered to the perpetrators of the act. Not very long afterwards other murders occurred in an equally mysterious manner, which threw the whole country into a state of excitement. Afterwards a small party was attacked by some unknown Indians, a portion killed and some left to tell the tale. The tracks of the strangers were followed through the woods, always keeping to the east, until they were lost. Following this clue, the chief of the Bri-bris sent out a party of armed scouts, who climbed to the summit of the dividing ridge, overlooking the Tilorio. From here they discovered for the first time that they had neighbors; seeing their houses and cornfields in the distance. A large war party was fitted out; they passed the mountains, and without warning descended on the unsuspecting enemy, killing large numbers. After this a desultory warfare was kept up; each party striving to take the other unawares, and to capture as many heads as possible. This went on until the Tiribi, reduced to a handful, sued for peace and submitted as a conquered people to the Bri-bris. Since then, the chief of the Bri-bris has always retained the right of final choice of chief of the Tiribis, after nomination of the candidate by his own people. Beyond this, no actual control has ever been exercised. The Tiribi story does not differ from the above, except in the origin. It throws the blame of the first aggression on the Bri-bris. In some respects the Tiribis are superior to the Bri-bris. The children are more respectful to their parents; the women are more modest in dress and behavior, and the men are more industrious. This is their boast, and while they look down on the Bri-bris, the latter despise them as a conquered people. Very little communication occurs between the two tribes, and I could learn of but two cases of intermarriage between them.

I have already said the Tiribis and Cabecars are under the political rule of the Bri-bris. The form of government is extremely simple. One family holds the hereditary right of chieftainship, and up to 1873 the reigning chief had theoretically full powers of government. The succession is not in direct line, but on the death of the incumbent, the most eligible member of the royal family is selected to fill the vacancy. Often a son is passed over in favor of a second cousin of the last chief. The present chief is first cousin of his predecessor, who was nephew of his predecessor, who was in turn a cousin to his.

Formerly the chiefs held only a nominal control over their people. The principal advantages derived from the position were rather of a social than a political nature. The chief was conducted to the best hammock for a seat on entering a house. He was treated to their great luxury, chocolate, when persons of less note were fain to be content with chicha. But in case of a quarrel the chief had to defend himself from the blows of the long, heavy fighting-stick like any ordinary mortal. Within the last decade or two, the traders, by throwing their influence on the side of the chief, have caused him to be treated with more respect, and endowed him with the attributes of a judge over his people, in all ordinary disputes. About 1870 or 1871, Santiago, the then chief, paid a visit to Cartago and San José; was well treated, and received an appointment from the Government, for the position which he already held, with the full approval of his tribe. It had been customary for the heir-apparent, the future successor, to hold a position as second, or subordinate chief, with little or no authority. One Lapiz was at that time second chief, and claimed that he was more entitled than the other to the chieftainship. Exaggerated ideas of great mineral wealth in "Talamanca" have been long held in Costa Rica and the Commandant of Moen, a little settlement on the Atlantic coast, used principally as a penal station, conspired with Lapiz against Santiago. This individual, named Marchena, advised Lapiz to assassinate his chief, and thereby place himself at the head of the tribe.

It seems that Marchena's plan was to put a creature of his own over the Indians, so as to gain access to the supposed rich mines and thereby benefit himself. Instigated by a "Christian," the savage, nothing loth, conspired with his people, but Santiago learned of it and made efforts to arrest him. Learning of this, he fled to the mountain fastnesses of Bri-bri where, broken down by disease and hardships he died, leaving, Indian like, his revenge as a legacy to his adherents. Santiago, who was a drunkard and, when intoxicated, a tyrant, gradually enstranged his people from him, and his relatives, Birche and Willie, placed themselves at the head of the opposition. The occasion sought for was not long in being found, and one morning Santiago was shot in the woods by an ambushed party, who at once took possession of the government, burnt their victim's house, appropriated his effects, including his three wives, and defied his friends. Birche, as the oldest of the two cousins and claimants to the chieftainship, took precedence and Willie became second chief. Mr. John H. Lyon, an American from Baltimore, who had lived in the country since 1858, had acted as secretary to Santiago, and only their respect for an upright man who had always treated them justly, coupled with the fact that he was not a "Spaniard," prevented them from venting their resentment on him, in common with the other friends of the murdered man. He remained at his house for some weeks despite the storm. But at last, thinking discretion the better part of valor, he left the country with his Indian family and remained absent some months. On his return he found matters settled after a fashion: the Birche party in power, but by no means secure against an outbreak from the friends of Santiago, who only wanted a leader. They urged Lyon to head them but his better council prevailed, and they perforce accepted the situation. I visited the country first in March, 1873. accompanied by the Commandante of Limon, Don Federico Fernandez. He then formally approved of Birche as chief, Willie as second, and re-appointed Lyon as Secretary. This was a great step in advance for Birche who now, for the first time, felt himself secure. The assassination of Santiago was practically ignored, but they were told "to be good and not do it again." This was succeeded by an infinite number of petty quarrels between the two chiefs; each disliking the other, and each wishing the other out of the way. By dint of constant interference on the part of the foreigners, they were prevented from coming into actual collision, although one attempt was made by the friends of Willie to kill Birche, Lyon, myself and my assistants at a blow by planting an ambush for us on one of our journeys. However, in December, 1873, business taking me to San José, I induced Birche to accompany me. On my advice, Don Vicente Herrera, the Minister of Interior, gave to Birche a formal commission as "Jefe Politico" of Talamanca, confirmed Willie as second chief, and appointed Mr. Lyon "Secretary and Director of the tribes," fixing suitable salaries for each. This was the first time that the tribe had formally submitted to the Costa Rican government. The action of Santiago was purely an individual affair, and looked on with great disfavor by the tribe. Matters went on very well for a few months under the new regime. But Birche, a man of little capacity, at the same time a coward and a tyrant, could not be content with his position. He began a system of ill treatment against which the people grumbled, but which they feared to resent. At first both Lyon and myself tried to quiet the complaints, believing that punishment had been justly inflicted, and knowing that

"No man e'er felt the halter draw
With just opinion of the law,
Or held with judgment orthodox
His love of justice in the stocks."

But it soon became apparent that his majesty (they are always called king) was abusing his power. The Indians dared not quarrel with Birche, for fear of offending the government, but came to Lyon almost daily with complaints. At last we decided to effect a change. Birche went to Limon to draw his salary, and at the same time to complain of a purely personal quarrel with Willie, in which he had fared worst. I arrived there a few days later, having completed my exploration, and being on my way to the Capital. On being asked for information and advice by the Commandante, I told the story and urged his removal. This could only be done by the minister, but he was suspended until the decision of that officer could be obtained. In a few days I saw Mr. Herrera, and after a conversation he decided to endorse the Commandante's action. Birche was accordingly removed, Willie was given a nominal chieftainship, and Lyon instructed to assume all responsibilities. Thus in less than two years the people have, without knowing how it happened, been deprived of their hereditary chiefs, and a foreigner placed over them. Willie remains with the empty title of chief without even the power to issue an order or punish an offender, except when ordered by Lyon. This gentleman has their entire confidence and respect, and many of the Indians begged to have even the title taken away permanently from the "royal" family. I have been thus prolix on this branch of the subject, because I was an eye witness, a participator, in the latter part of the events I relate. Trivial as they are, they may interest some, throwing light on the manner in which one tribe after another is subdued.

A strange fatality seems to hang over these Isthmian Indians. Even when not brought into contact with the debasing influences of civilization, the tribes are visibly diminishing. Less than two centuries ago, the population of Talamanca, as Costa Rica calls her southeastern province, was counted by thousands, now barely 1200 souls can be found. The Shelaba tribe is extinct; the Chanquinas are at the point of extermination, the Tiribis number but one hundred and three souls, and Lyon tells me that the Cabecars of the Coen have diminished fully one-half within the last seventeen years, while the decrease in the Bri-bris is hardly less rapid.

During my travels in Talamanca I collected in each district an accurate enumeration of the population. My process was to get together several of the most intelligent and well-informed men in the district; cause them to compare notes and then to tie a series of knots in strings as they are accustomed to do; different kinds of knots distinguishing the sexes. Each house was counted separately, so that I obtained an exact census of the whole country with the following results. This cord census is now in the museum of the Smithsonian Institution, with many other articles, illustrating the life and customs of the people.

The population of each district is as follows:

Tiribi	103
Uren	604
Bri-bri	172
Cabecar	128
The Valley	219
Total	1226

This covers all of the water-sheds of the Tilorio and Tiliri rivers except two small bands; the Changinas on the Changina branch of the Tilorio and a refugee remnant of the Cabecars on the extreme head of the Tiliri. Probably an additional hundred would cover all of these.

On the North or Estrella river, and on the Chiripo, there are a few more Cabecars who have little communication with the headquarters of the tribe, but who are in the habit of going out to Limon or Matina for what little trade they require. These are probably in all, not more than 200 or 300 in number. Nearly all speak Spanish and they are gradually approximating to civilized or semi-civilized ways.

The cause of the rapid decrease in the population is their extreme indolence. With a country fitted to produce all the fruits of the tropics; where maize grows luxuriantly, and where cattle and pigs increase without care or labor; they are content to make plantains their staple, and almost their only food. Chicha the form in which most of their maize is used, is a beverage very slightly intoxicating, if drank in large quantities, but the amount of nutriment derived from it is unimportant. Meat, whether of domestic or wild animals, is a rarity and a luxury, and the banana or plantain make up all deficiencies. The natural consequence of a bulky and comparatively innutritious diet is a low physical state. The system has little resisting power against disease, or healing power over wounds. A slight attack of coast fever, which, with an ordinary strong man of our own race, would be comparatively harmless, is very apt to terminate fatally with these people. Indolent ulcers are so common that perhaps a full fourth of not only adults, but even children have them, usually on the legs, originating in some slight scratch or bruise; and very few of the elderly persons are without their scars. These ulcers often last for years, and I have seen them as broad as the two hands opened side by side. Although the local diseases are few, the entire absence of medical treatment, the ignorance of the first principles of hygiene, and the universal negligence of the sick, on the part of the well, all contribute to shorten the average life-term of the people, so that very few old men or women are to be found, and the mortality is so great among the young that the deaths more than counterbalance the births. Unless some great change takes place, the whole of the tribes of Talamanca will have disappeared within two or three generations more. The Tiribis, who like the others have strict rules about marriage, within certain degrees of consanguinity, are now so reduced that several young men and women are to-day forced to remain unmarried for want of proper mates sufficiently removed in relationship. But at the beginning of this century they were powerful enough to give battle to the Bri-bris. The Changinas and Shelabas have disappeared and the fate of the other tribes requires no prophet to foretell.

Physically, the people of all the tribes bear a strong resemblance to each other. They are of short stature, broad shouldered, heavily built, full in the chest, with well-formed limbs, and well muscled throughout. Their color is similar to that of the North American Indians, or, if anything different, perhaps a little lighter. There seems to be but little, if any admixture of foreign blood among them. Their history would hardly lead us to expect it. They have lived very exclusively, and it has hardly been half a century since they have ceased to live in a state of open war with all intruders from the coast side. The Spanish occupation closed so disastrously over a century and a half ago, was of too short duration, and and the whites were too few, to make a permanent impression on a then populous country.

The following measurements taken from my servant, a full grown man, who is not more than an inch, if so much, under the average height, will give a fair idea of their build. He measures in height, 5 ft. 1½ in., circumference of chest, under the arms 35¾ inches; of hips 34 inches, of waist 33¼ inches, length from axilla to tips of the fingers, 24½ inches; leg, from the groin to the ground, 29 inches. Both sexes are marked by an almost perfect absence of hair from all parts of the person except the head; where there is a dense growth of coarse, straight black hair. This the women plait with considerable taste. The men wear theirs cut moderately long and of an even length all round; or a few retaining an older fashion, have it a little over a foot long, apparently its entire natural length, and either let it stream loosely over the shoulders, gather it into two plaits, or twist it into a roll, bound with a strip of mastate, and coiled at the back of the head in a round flat mass.

The breasts of the women are not conical, as occurs with many, if not most of the Indian races; but are fully as globular as those of the European or African. Nor are they directed laterally. They are not generally large, though some marked exceptions occur to this rule. But they have one strongly marked peculiarity. The entire areolar area is developed into a globular protuberance, completely enveloping and hiding the nipple. The development of this part begins with, almost before, that of the mammary gland proper, on the approach of puberty, and is more obvious then, than after the gland has acquired its full rotundity. After marriage, the areola gradually sinks, leaving the nipple standing out prominently in its centre.

In treating of the manners and customs of these people, I shall include the three tribes of Tiribi, Bri-bri, and Cabecar as one, and shall only mention them separately where points of difference occur. First in the order comes the birth of the young savage.

All the world, or rather all the ignorant world, and even a part of that which considers itself reasonably enlightened, entertains a belief in the influence on the child, of certain impressions made on the mother during pregnancy. Doubtless the general mental state of the mother has an influence on her progeny. But the belief exists among these Indians, in its full force, that the sight of certain objects by the mother will influence her child physically. They go further. The mother is given to wearing certain charms to that end. The eyes of the fish hawk give the future fisher the power to see his prey beneath the water; the teeth of the tiger (also worn by both sexes for purely ornamental purposes), when used as an amulet makes the future hunter swift and strong in the chase; the hairs of a horse make him strong to carry loads, and a piece of cotton pushed inside of her girdle by a white man, is certain to make the child of a lighter complexion.

When the time of parturition approaches, the father goes into the woods and builds a little shed, at a safe distance from the house. To this the woman retires as soon as she feels the labor pains coming on. Here, alone and unassisted, she brings forth her young. Difficult delivery is as rare as among the lower animals. As soon as the delivery is effected, the mother of the woman, if present, and in her absence, some other old woman approaches the mother and, with great circumspection to avoid the defilement of *bu-ku-ru'*, of which I shall speak further on, places within her reach a piece of wild cane, so split as to make a rude knife. The mother ties the umbilical cord and severs it with this knife. No other kind is permitted. She is also supplied in the same manner with some tepid water in a folded plantain leaf, in which she washes the child. She then collects the after-birth, &c., and buries it, after which she goes to the nearest water and bathes herself. An *awa*, or medicine man then appears on the scene. He causes the mother to theoretically wash herself again, by dipping her fingers into a calabash of water, which he forthwith drinks. He then lights a pipe of tobacco, blowing the smoke over her. He then purifies himself by washing his hands, after which, and not before, all are permitted to return to the house. The recovery of the mother is so prompt that it may be more properly said, she has nothing to recover from. I have seen a young mother, with her first child not yet a week old, attending to her ordinary duties as if nothing had happened.

The matter of names is very loose and arbitrary. It is almost impossible for a stranger to learn the true name of an Indian, directly from the person himself, although his friends may divulge it, and this is looked upon almost in the light of either a breach of confidence, or a practical joke. After long acquaintance, they may be prevailed upon, but even then are more apt to give a false name than to tell the truth, so great is their reluctance. One fellow, who was my servant for over three months, after always denying having a name, at last told me a pet name, or "nick-name" that he had had as a child. It is customary for children to have provisional names, or to be called only "boy" or "girl" as the case may be, until the whim of an acquaintance or some equally arbitrary circumstance fixes a title to them. Besides the native name, generally derived from some personal quality, or not seldom the name of some animal or plant, almost all of the Indians possess a foreign name, by which they are known, and which they do not hesitate to communicate. Among themselves, when the name is unknown, a person is called by the name of the place where he lives. Mr. Lyon says all the women have names, as well as the men. But my experience with them is never to have heard them called by other titles than "girl," "woman," "*wishy*" (applied familiarly to young married women), or "so-and-so's wife" or daughter, except in the case of a few of the more civilized men, who have given Christian names to their families.

Children are not generally weaned early. In case of the birth of a second child, the first is weaned perforce. But it is nothing strange to see a child well able to walk, say even two years old, go to the breast as a matter of course, although sufficiently accustomed to more solid food.

Small babies are carried on the back, astride the hips of the woman, and supported by a broad strip of bark or cotton cloth, passed around both, and secured in front by a dexterous tucking in of the ends. When they become larger, they are carried on one hip, supported by the arm; or are placed on top of the load, if the mother is traveling. They sit perched on the bundle, with a foot dangling either over or behind each shoulder of the mother, and soon learn to hold on like monkeys.

The training of the youth is left almost entirely to themselves. Among the Tiribi they are taught to respect and obey their parents, but in the other tribes they are more insolent and disrespectful to their parents than to other persons. I have seen a boy of ten years old absolutely refuse to obey some trifling command of his mother, and she seemed to have no power to enforce her order. The little girls learn early to accompany the older girls and women when they go out to bring water. Their usual station, in the house, is at the side of the fire, where, as soon as they are large enough, they assist in fanning the fire, preparing plantains for the pot or watching the cooking. The boys will sometimes deign to hunt fire-wood, but they are more apt to be playing by the side of the river with mimic bow and arrow, learning to shoot fish under water. Their toys are mostly diminutive copies of the tools and weapons of more advanced age. The machete of the man is represented by a good sized knife, often the only article worn by the boy; the long hunting and fishing bow is foreshadowed by one a yard long, perhaps made of a simple piece of wild cane; the blow gun, a tube longer than the person, is in constant use; and I have seen some few actual toys such as a top made of a large round seed with a stick through it; and a rattle differing only in the degree of care in the making, from those used by the priests in their incantations.

The arrival of puberty is the signal for marriage, at least on the part of the girls. The courtships, if such they can be called, are carried on principally at the chicha drinkings, and I am assured that very few young women retain their virginity until marriage. A plurality of wives is allowed at the option of the husband. Many have two, and some three women. When a young man wishes to marry, having arranged with the girl, he applies to the father. The consent is practically a foregone conclusion; but the details of the bargain must be arranged. In most cases, the groom goes to live at the house of his father-in-law, becomes, at least for a time, a member of the family, and contributes with his labor to the common support. Girls are thus available property to their families. But in case the man already has a wife; is in short, settled in life, and has his own home, he may not want to change his residence. He then compounds with the family; giving a cow, a couple of pigs, or other equivalent for the woman, in place of his services. No form of ceremony is required, and the marriage lasts as long as it suits the convenience of the parties. In case of infidelity on the part of the woman, or undue cruelty on the man's part, they may separate. Sometimes, if the woman is unfaithful, the man whips her severely, and perhaps returns her to her family, or she, in a fit of resentment, leaves him. This may be for a year or so, or may be final; but during such separation either party is at liberty to make new connections, thereby remaining permanently apart.

Probably there is no better place to mention kissing than in connection with courtships and marriages. This agreeable custom seems to be entirely unknown. I have never seen one person among them kiss another, not even a mother her child.

There are certain limits within which parties may not marry. The tribes are divided into families, or something analogous to clans. Two persons of the same clan cannot marry. This is now a source of difficulty among the Tiribis. The tribe is so reduced that a number of marriageable persons of both sexes are unable to find eligible mates. I could not ascertain exactly how the question is settled as to which clan a person belongs, whether he inherits from father or mother, but so far as I could gather, I think from the father. Cousins, even to a remote degree, are called brother and sister, and are most strictly prohibited from intermarriage. The law, or custom, is not an introduced one, but one handed down from remote times. The penalty for its violation was originally very severe; nothing less than the burial alive of both parties. This penalty was not only enforced against improper marriage, but even against illicit intercourse on the part of persons within the forbidden limits. Mr. Lyon related to me a case that occurred since he has been living in the country, where the power of the Chief Chirimo was insufficient to protect a man who married his second or third cousin. Fortunately for the delinquents, they succeeded in making their escape, though with difficulty, being followed two or three days' journey by the avengers.

Infidelity is not rare, and the husband has the redress of whipping the woman and dismissing her if he desires, and of whipping her paramour if he is able. But so cautious are the people about the blood limit of intermarriage, that a woman on giving birth to an illegitimate child, for fear that it will not know the family to which it belongs, will usually brave the punishment, and at once confess its paternity.

As cousins are called brother and sister, so are not only the brothers and sisters, but even the cousins of a wife or husband all called indiscriminately brother and sister-in-law; so that a person may on a single marriage find that he has annexed fifty or a hundred of these interesting relations.

On the death of the head of the family, the next oldest brother, or in default of a brother, a cousin or uncle assumes his place, and is then called father by the children. This does not involve any especial material duties, such as the support of the family; but is rather a sort of honorary title; giving him, however, the ruling voice in any family council or discussion.

On the death of an individual; if a young person, a woman, or a person of but little consequence, the body is prepared as soon as possible in the manner described below, and carried to the forest; but if a person of more consideration, there are some preliminary ceremonies. These I had the opportunity of witnessing in the case of an old man who died on the Uren when I was present. He belonged to one of the distinguished families, an ancestor, perhaps his father, having been one of the leaders in the war with Tiribi, and he the heir to, and possessor of, one of the few gold "eagles," or insignia of rank. He died in the night, and next morning, the body being in his hammock, covered with a piece of bark cloth; all of the chicha, chocolate, and food that the poor people of the house could get together on short notice were prepared. A fire was lighted, amidst singing, by twirling a pointed stick in a socket on the face of another. This was the sacred fire, which was communicated to a small heap of wood placed on one side in the house. This could be used for no common purpose whatever. No ordinary fire could be lighted from it; not even could one use a stick of it to light his pipe. It must burn continuously for nine days. In case of its accidentally going out before that time, it must be relighted in the same manner as at first; and at the end of that time, only a priest could extinguish it, and he only with a calabash of chocolate, and during, or at the end rather, of the suitable incantation.

The custom of burying or otherwise placing with the dead all of his valuables, evidently existed at one time with these people. The Tiribis, who bury their dead, did so, up to within the memory of persons still living, and all matters that could not be buried, like live stock, fruit trees, &c., were ruthlessly destroyed. A more practical method has grown up with the present generation, and they now divide the property of the defunct among the heirs, with as much avidity as in more enlightened communities. So do the Bri-bris and Cabecars, but these compound with their consciences. Whether the Teribis have a similar custom, I am not prepared to say, not having seen a funeral, and having no information that I consider sufficiently trustworthy.

The next step after lighting the fire, was for the master of ceremonies, appointed by mutual consent, to cause to be collected some small scrapings of a peculiar wood, called Palo Cacique by the Spaniards. It is a wood used only for walking sticks, and will be again referred to in that connection. He also obtained a large lump of cotton wool, some seeds of a species of pumpkin, and a small root of sweet yucca. All the male friends of the deceased present, seated themselves on low benches in a double line, facing each other, with another bench between. A part of the cotton, spread out so as to make a bulk about the size of a man's hand, was placed in front of the principal person, who then began in a sing-song tone between a recitation and a chant, to relate the merits and deeds of their departed brother. As he proceeded, and mentioned for instance that he had planted much corn, he laid carefully on the cotton a piece of shaving which he said was the "planting stick" used in that operation. Another laid aside of it a piece of pumpkin seed, which represented the corn. Another taking up the song, related how he had shot fish, and another shaving was the arrow. An impromptu string a couple of inches long, twisted out of the cotton, and stained red with the powder from some annatto seeds, was a rope with which he had led a cow, bought years before in Terraba. This lasted for an hour, until every tool or weapon he had ever used was represented by a little pile of seeds and shavings on the cotton. But he was a great man and his "eagle" was not to be forgotten. A very rude imitation of it was cut out of the skin of the yucca root and placed on top of all his other property, and then the edges of the cotton were doubled over making all into a ball. This was placed on his breast, next his body, and he was thus armed and equipped with all he had used or owned in this world, ready for use in the other; and his heirs none the poorer.

The body was then enveloped in the piece of "mastate" or bark cloth that he had used as a blanket, together with the hammock in which he swung. A quantity of "platanillo" leaves, a leaf not unlike that of the plantain, but only half the size and much tougher, were placed on the ground, two or three deep. The bundle was laid on this, the edges of the leaf envelope, doubled over, and dexterously tied by strips of bark string and the whole turned out a very respectable Egyptian mummy done in green. By means of three strings, this was swung under a pole, ten feet long, raised on the shoulders of two men, who trotted off unconcernedly to the woods a mile or so distant. They were accompanied by two or three more, armed with machetes.

A little boy whom I had for a servant for a few months, died on one of my journeys. We watched by him and did all in our power to save him, and were assisted by two of our men, one of whom was an "*awa*" or doctor. As soon as we saw that he was dying, and I had given up the last hope, the *awa* took charge. He motioned us all off. From that moment the moribund becomes unclean and only the *awa* can touch him. As soon as we pronounced him dead, the doctor covered him up. Next morning, the death taking place about midnight, without ceremony he was bundled up in his blanket and the usual leaves, and carried off in the same manner to the bush. But he was of no consequence. Only a boy who was nobody and had done nothing. I mention this case to show the difference in treatment, according to the person.

Next to a woman in her first pregnancy, the most *bu-ku-ru'* (unclean) thing is a corpse. An animal that passes near one after it is placed in its temporary resting-place, is defiled forever, and must be killed, as unfit for food. Accordingly, an unfrequented spot is selected, where tame pigs or cattle never go. Here a low bench is made of straight sticks, about the size of a coffin, raised a foot or two from the ground; it is carefully fenced in; the corpse is laid on it, and the whole is then covered with another horizontal layer, making a sort of box, carefully bound together with vines. Over all, a pile of branches and brush-wood is thrown so that buzzards and other carrion-eating animals cannot obtain access to the body. The body remains here about a year, to allow complete decomposition.

In the meantime, the family, or next of kin, on whom devolves the responsibility, proceeds to secure a sufficient number of animals, pigs, or beeves, according to the importance of the defunct. He also plants a corn-field, to supply the material for the chicha. About a year, more or less, after the death, one or more priests are engaged. Generally one is sought and he selects his assistants. For an ordinary person, one is sufficient; while for a chief, or person of distinction half-a-dozen are hardly enough. The chief fixes the time when he will be ready. Another official, a steward, called *Bi-ka'-kra* is also engaged. This latter personage takes entire charge as commissary and master of ceremonies. Under his direction, the corn is ground for the chicha. The number of bunches of plantains that he orders, is obtained; the animals are killed and cooked as he directs; and the food and drink are served to whom, and in what quantities he designates. The host resigns all to him and becomes thenceforth merely a guest, until all is over.

When the day approaches, a party goes to the place where the body was deposited. One person, set apart for similar unclean work, opens the package, cleans and re-arranges the bones and does them all up in a bundle about two feet long; enveloped in a piece of cloth of native make, prepared by being painted in an allegorical manner.

These cloths, about four feet long by two wide, are painted with a red vegetable juice, in figures two to four inches long. The devices vary according to the cause of the death of the individual; whether it be from fever or other disease, old age, snake bite, wounds, &c. One of these cloths, in the Smithsonian museum, is painted for a person who is supposed to have died from snake bite.

The bones, having been tied up in the new bundle, are carried, again under a pole, to the house where the feast is to be held, and are there placed on a little rack overhead, out of the way of persons passing underneath.

Everything being ready, the first installment of food cooked, the chicha brewed, and chocolate boiled, the feast begins.

I had the rare good fortune, not only to witness the ceremony at the death of the persons mentioned above, but also to be present at the death feast of the chief Santiago. That is to say, I saw all that happened on the first and the last days; the intervening thirteen or fourteen being all alike; a succession of eating, drinking, dancing; a disgusting scene of carousal and debauch that did not possess even the merit of variety.

The feast was held in a large house, adjoining the residence of the chief Birche. The house is about seventy-five feet long and forty wide; the ends being round, and the only light entering by the large doorway left open at one end. A little rack, made of wild cane was tied up to the sloping side of the house, about eight feet from the floor, and on this was laid the bundle containing the disjointed skeleton of the murdered chief. At a given signal, the principal singer or priest took his position on a low stool, flanked by the other priests and some volunteers. All were regaled with chocolate served in little gourds. The priest began a low chant and two men started twirling the stick to light the fire. As fast as one tired, another took his place until the sparks glowed in the pit bored in the lower stick. A yell from the priest announced this, and a piece of cotton wool was ignited from the burning dust; with this the fire-wood, previously prepared, was lighted and the fire placed under the remains. Here it was kept up until the end of the feast. After the lighting of the fire, singing and dancing began in earnest, interrupted occasionally by eating and drinking.

The dances are very similar; the principal differences visible to an observer are in the disposition of the dancers, whether in a circle or in one or two straight lines. In the latter case, the two lines are parallel, and the dancers face each other. The dancing is kept up to the "music" of small drums, carved out of a solid piece of wood, with a single head, made of the belly skin of the iguana; the other end is open. The drum is held under the left arm, and is beaten with the tips of the fingers of the right hand. The drummers, ranged in a line, sing a monotonous song, with a chorus; the time being beaten on the drums. Sometimes a dried armadillo skin is scraped with a large bean-like seed ; in the same manner as I have seen the negroes of the West Indies scrape a roughened calabash with a bone. The dancers clasp each other over the shoulder, around the waist, or hook arms ; both sexes taking part in the dancing, but not in the singing or drumming, these being the especial province of the men. The steps are usually about three forward and to one side, and then the same number backward. When arranged in a circle, this carries them gradually around the musicians. When in a straight line, they keep on the same spot. The songs are a sort of recitative, sometimes impromptu, sometimes of fixed words ; the chorus a sort of "fol-de-rol," a series of meaningless syllables. These songs for dancing must not, however, be confounded with the sacred songs of the priests, of which I shall have occasion to make fuller mention in the proper place.

The dances are kept up nearly all, and sometimes all night at the funeral feast ; the participants retiring from time to time and sleeping an hour or two when exhausted, and returning with renewed vigor to chicha drinking, eating, and dancing. It is particularly on these occasions, when the older people are too drunk, or too busy to keep strict watch, that the younger folks manage to evade their vigilance and ——. These eminently practical courtships almost invariably precede the asking of the father's consent by the would-be bridegroom.

After more than two weeks of this license and debauchery, during which three cows, about a dozen pigs, hundreds of bunches of plantains, several quintals of rice, and hundreds of gallons of chicha had been devoured, the *bi-ka-kra* or steward announced that the commissary had given out and the riot must come to an end. I was notified according to previous agreement and went at the time appointed. As distinguished guests, our party of four were shown to the best hammocks, where we were seated, and in a few minutes served with cups of chocolate. In a little while, all of the priests seated themselves on low benches, the leader in the middle. The lay chorus singers were ranged in a double line facing each other and below the priests. The fire was carefully carried from its place under the corpse and piled almost between the feet of the principal priest. All drank chocolate and the priests sounded their rattles. The leader began a low dirge-like song in the sacred jargon, which I was told described in detail the journey of the defunct to the other world. It told of the dangerous rivers he had to cross, where alligators lay in wait to devour him ; of the great serpents who disputed his path ; of the high hills he had to climb with weary steps ; of the fearful precipices he must scale ; of the beautiful birds with sweet songs, compared with which even the flute-like *silguero* was as a crow ; of the gorgeous butterflies that lightened up the path like flying flowers, and finally of his safe arrival at the country of the great *Si-bu*, where he would have nothing to do but eat, drink, sleep, and enjoy himself.

The song was divided into stanzas, and the priests all followed the lead of their chief, the words being a series of set phrases, but in a language in part unintelligible to the uninitiated. At the end of each stanza was a chorus, where the priests, who during the stanza kept time with their rattles, now gave a peculiar twirl, and the lay singers joined in the chorus.

As the song approached its end, the leader was furnished with a big gourd of steaming chocolate, holding about a quart. As he finished, landing the dear departed safe beyond further troubles, he announced it with a most unearthly yell, in which all hands joined; he at the same time turning the chocolate over the fire, totally extinguishing it. The party at once arose and for a minute or two all was bustle and confusion of preparation.

A person, whose office it is to handle the dead, endeavored to lower the bundle but it was a little out of his reach. Nobody else could touch it for fear of defiling himself. To lend a hand would have cost an Indian three days of purification. I drew my long knife, which all learn to carry in this country, as an actual necessity; and with a couple of blows cut the fastenings and brought the little cane rack, bundle of bones and all, tumbling into the outstretched arms of the official, with much more haste than solemnity. Nobody seemed shocked, and being a foreigner, and withal a medicine man, who had made cures where their best doctors had failed, I was of course impregnable to *bu-ku-ru'*. The aforesaid official now lashed the package to its stick, and two long slender strings of loosely twisted cotton were tied to the head of the package.

Santiago had had three wives. One of them had re-married to his successor; but there were two remaining in widowhood. A procession was formed. First came the priests with their rattles. Next the chorus singers with their drums. Next the corpse, borne by two men, and preceded by the two widows, each holding the end of one of the cotton strings, leading the dead, as it were, to his final resting-place. Next ourselves, as the most distinguished persons present, and escorted by the two chiefs. Behind us came the older men, and following them the usual rag, tag, and bob-tail of young men, women, boys, and persons of no account generally. Some of the boys however, true to boy nature, were as usual irrepressible, and instead of keeping decorously in place, skirmished ahead, and on the flanks of the procession, mounting stumps, logs, or other commanding points to take in the general effect of the pageant. As the procession filed out of the house, some old chicha jars were carried out and ostentatiously broken; but I observed that nothing of real value was destroyed. As soon as the line got fairly under way, the priests struck up another song which was kept up until the procession halted.

Everybody had been on so long a debauch that it was decided to take a rest of three or four days before the party started off. But it was necessary that the bones should be removed from the house. A temporary ranch had therefore been built a few hundred yards distant; and to this the remains were carried and deposited until the bearers were in a fit condition to proceed.

The final disposal of the remains is a matter of great care. The whole of the tribe goes to the district of Bri-bri for this purpose. The receptacle is a square pit, about four feet deep and ten feet square. This is paved on the bottom with stones, and is roofed over from the weather, by a series of heavy hewn slabs of a very durable wood, open on the front and ends, and sloping to the ground at the back. Each family possesses one of these pits and here, after the funeral feast, the bundle of bones is carried and deposited. After the rest, the remains of Santiago were carried to the "royal" pit and deposited without further ceremony.

The Cabecars, according to Mr. Lyon, have about the same ceremony, but their pits are mere holes, not paved, and covered by planks laid on the ground level.

The Tiribis have a death feast, but it differs in some respects from the others. The body is buried immediately after death, but no longer with the property of the deceased, and, of course, the defunct is not present at his final feast, as with the Bri-bris.

Mr. Lyon, to whom I owe much of the information in the present memoir, has described to me one circumstance, in connection with these death feasts, that I have not witnessed. The warriors among the Bri-bris, who fought in the war with the Tiribis were honored with a little different ceremonial. They are now all gone, and the ceremony is extinct. At the death feast, a person entered, clad in a long gown, wig, and mask. The gown and wig were made of mastate, or bark cloth, covered with "old man's beard" moss, sewed all over it, making a shaggy and nearly shapeless mass. The mask was made of half a "tree calabash," properly fixed up with a wax nose, &c. A copy of this entire dress was made for me by an old Indian, and is now in the Smithsonian museum. The person thus accoutred, took part in the dance, made free with the women and scared the children without let or hindrance. Mothers with young children took them to him and placed them for a moment on his shoulder, "to prevent the evil spirit from doing them harm." Neither Lyon nor the Indians could give me a very clear account of what spirit, whether good or evil, this represented. But the people seemed to regard him as rather of the malevolent sort; to be classed under the general head of "*Bi*" or Devil. Doubtless this, at one time had a distinct meaning, now lost.

No strictly religious belief can be said to exist among these Indians, in the sense that it is usually understood among us. They have, however, a series of ideas or beliefs which affect their daily lives and are never lost sight of. In connection with the funeral feast, described above, I have referred to their idea of a future state.

During the year that the body lies in the woods, the disembodied spirit prowls around, living on wild fruits, of which the wild cacao is the only one of which I know the name, although others were also pointed out to me. At the end of that time, when the funeral fire is kindled, the spirit is thus attracted to the feast, whence it departs on its final journey.

When I asked an Indian where it went, he responded, to the country of *Si-bu'*, and in reply to the question; where is that? he pointed unhesitatingly to the zenith. On inquiring where the road was, he told me it was invisible to the eyes of the living, but that the spirit (*wiy'bru*) could see it.

In the other world there are no troubles, no cares. There is plenty to eat and to drink, of those things that the Indian loves most here. Plantains and corn are never wanting; meat and chicha are always to be had; and chocolate, the luxury, *par excellence* of the Costa Rican Indian never runs out, or becomes scarce as, alas too often, it does in Talamanca. He needs all his arms and implements, but it does not seem that he will be obliged to work. These little discrepancies, the wisest *Tsu-gur* does not attempt to explain. After death, the soul remains wandering about near the corpse until the burial feast. Then, by means of the songs of the *Tsu-gurs* or priests, it makes its journey to the "promised land."

Their superstitions are however, somewhat more definite and tangible since they affect their every day actions. There are two classes of uncleanness, *nya* and *bu-ku-ru'*. Anything that is essentially filthy, or that was connected with the death of a person is "*nya*," anything unclean in the Hebraic or Hindu sense is *bu-ku-ru'*. But *bu-ku-ru'* is even more powerful than it is supposed to be by the Orientals. It suffices not only to make one sick, but even kills. In a party where *bu-ku-ru'* is excited, it does not affect all alike, but only attacks the weakest. *Bu-ku-ru'* emanates in a variety of ways; arms, utensils, even houses become affected by it after long disuse and before they can be used again must be purified. In the case of portable objects left undisturbed for a long time, the custom is to beat them with a stick before touching them. I have seen a woman take a long walking stick and beat a basket hanging from the roof of a house by a cord. On asking what that was for, I was told that the basket contained her treasures, that she would probably want to take something out the next day and that she was driving off the *bu-ku-ru'*. A house long unused must be swept and then the person who is purifying it must take a stick and beat not only the movable objects, but the beds, posts, and in short, every accessible part of the interior. The next day it is fit for occupation. A place not visited for a long time or reached for the first time, is *bu-ku-ru'*. On our return from the ascent of Pico Blanco, nearly all the party suffered from little calenturas, the result of extraordinary exposure to wet and cold and of want of food. The Indians said that the peak was especially *bu-ku-ru'*, since nobody had ever been on it before. Even we foreigners were sick from it, and had any of them gone to the summit, they would have surely died. On one occasion, while buying some implements, I pulled down off a rack, two or three "blow guns" that, from the dust on them, must have lain there undisturbed for weeks, perhaps months. As I reached out my hand, I heared the warning cry of "*bu-ku-ru'*" from all around; laughingly disregarding it, and telling them that *bu-ku-ru'* couldn't hurt us, I began examining them.

Some of the people looked very serious and shaking their heads, said I would see before long, that somebody would pay for it. Two or three weeks after, a fine little Indian boy whom I had with me as a servant, poisoned himself by eating excessively of a kind of wild almond called variously the "bri-bri," or "eboe" nut. There was not an Indian in that party but who firmly believed that it was the *bu-ku-ru'* of the blow-guns that killed him. From all the foregoing, it would seem that *bu-ku ru'* is a sort of evil spirit that takes possession of the objects, and resents being disturbed; but I have never been able to learn from the Indians that they consider it so. They seem to think of it as a property the object acquires. But the worst *bu-ku-ru'* of all, is that of a young woman in her first pregnancy. She infects the whole neighborhood. Persons going from the house where she lives, carry the infection with them to a distance, and all the deaths or other serious misfortunes in the vicinity are laid to her charge. In the old times, when the savage laws and customs were in full force, it was not an uncommon thing for the husband of such a woman to be obliged to pay damages for casualties thus caused by his unfortunate wife. *Nya* (literally filth) is a much less serious affair. As soon as the woman is delivered of her child, she ceases to be *bu-ku-ru'*, but becomes *nya* and has to be purified in the manner already described. All the objects that have been in contact with a person just dead, are *nya* and must be either thrown away, destroyed, or purified by a 'doctor.' He can handle them, but must purify himself afterwards. The persons who assist in preparing the corpse, who carry it to the temporary resting-place, or who even accidentally touch it or the unclean things, are all *nya* and must be purified.

Purification from this latter uncleanness is a simple matter. The person washes his hands in a calabash of warm water, the "doctor" blows a few whiffs of tobacco-smoke over him, and the thing is done. But the former is much more serious. For three days the patient eats no salt in his food, drinks no chocolate, uses no tobacco, and if a married man, sleeps apart from his wife. At the expiration of that time, the warm water and tobacco smoke are called into requisition and the cleansing is complete.

Of Gods, deities, spirits, or devils, there are as follows; the "great spirit" or principal superhuman being is called *Si-bu'* by the Bri-bris and by the Cabecars; by the Tiribis he is called *Zi-bo'*, by the Terrabas *Zŭ-bo'* and by the Borucas, *Si'-bŭh*. A good spirit, from whom nothing is to be feared, he receives a sort of passive respect, but no adoration or worship. He is rather looked on as the chief of the good country, of the future state, but as not troubling himself much about mundane matters. It will be seen, therefore, that in their theology, the entire family of tribes is essentially monotheistic, although they have taken the first insensible step towards a plurality of gods, in the manner so admirably indicated by Max Müller, in his "Chips from a German Workshop." They believe in but one God, and assert his unity with an emphasis worthy of Moslems and yet their priests give him twenty names, in their songs. These names, so far as I could ascertain, all refer to his qualities. One Bri-bri, whom I had with me as a servant for over half a year, and from whom I obtained much valuable information, particularly in regard to the language, said to me, "Why do you foreigners ask us how many Gods there are? There is only one, and that is *Si-bu'*."

The Devil, or devils, are minor personages, who receive no worship of any kind. They are called, *Bi*, by the Bri-bris and Cabecars, *Au* in Tiribi, *Auh* in Terraba, and *Ka-gro'* in Boruca. The devil is generally malevolent, but does not seem to be specially feared. *Bi* among the Bribris is a term also used for a variety of lesser devils, or evil spirits who have special missions, like making people sick, &c. Some of these inhabit the less frequented parts of the forests and mountains, and are very jealous of their domains. People entering an unfrequented region, make as little noise as possible. If they make the local *Bi* angry with their noise, he will revenge himself by a shower or by causing somebody to fall and hurt himself, or to be bitten by a snake, &c. A person who has once been in these places can return with less risk, but all new-comers must keep at least a comparative silence. Another class of beings inhabit the rocks on the summits of certain mountain peaks. They live *inside* the rocks, not among them, consequently their habitations are undistinguishable to mortal eyes. They seem to have the same habits as ordinary humans. One of these peaks, a mile or two across a cañon, in front of a place called Sar-we, is thus inhabited according to the accounts of the people of Sar-we. They told me of hearing singing, the beating of drums, &c., coming from that direction. The configuration of the hills is such that a glance showed me, that a drum beaten at certain of the houses in the cañon of Uren, would echo back from this hill to Sar-we and thus account for the sounds. These people of the *U-jums*, as the naked peaks are called, are said to be the owners of the tapirs which roam through these solitudes. They are very jealous of their domains and cause, by some occult means, the death of any one who dares approach their homes. I could not induce an Indian to accompany us to the summit of Pico Blanco, partly on account of *bu-ku-ru'*, and perhaps more still for fear of the people of the *U-jum* or peak. In addition to these beliefs, they also believe in the efficacy of incantation by their *Awas* or doctors, of whom more immediately; and further in certain ceremonies or observances of their own. I have seen a woman carefully collect a bunch of some weed and taking it to the river wash her face, neck, breast, and arms with it. This was to bring good luck to the men who were at the time at work turning a stream to dry its bed, for the purpose of catching fish. She had her reward; hundreds of fish of 2 to 4 pounds weight were captured, and of a quality as fine as shad.

There is a peculiar wood, of which I shall have occasion to speak further on, used only for walking sticks for the chiefs and more eminent persons. The growing tree is unknown and it is only obtained by the accidental discovery at rare intervals, of a half-rotten trunk in the woods. It is prized principally for its color, which is between that of old mahogany and rosewood, and which is probably in part due to seasoning, or to some change in the heart, consequent on the decomposition of the surface. When an Indian finds one of these sticks, he marks the spot, but dares not take possession immediately. He must purify himself by a three days fast before he can begin work on it. It is believed that these sticks are under the protection of a poisonous snake, and if the person has not properly prepared himself, the guardian will revenge the outrage by biting him.

The privileged classes, apart from the chiefs, are three. Two of these are hereditary. The *U-se'-ka-ra* is a sort of high priest, and is of nearly as great importance in the eyes of the people as the chief. In fact, the time was, and not very long ago either, when the chiefs themselves made journeys to visit him as suppliants. The present incumbent is a youth of perhaps twenty-five years of age, and is not yet full fledged. His predecessor, his father, died recently, and, until after the funeral feast, he cannot enter fully into the exercise of his functions. The family lives far back in the hills of Cabecar, and, although a member of that despised tribe, has from time immemorial held undisputed sway over both it and the Bri-bris.

The former *U-se'-ka-ra* was very arrogant, and would hold no communication with foreigners. He claimed supernatural powers, and held frequent interviews with spirits. On these occasions he went alone to a cave, several miles from his house, and spent days together there. On his return he would not converse even with his own family. Nobody but his familiar, now a very old man, was allowed to serve him, or even to speak to him for a certain number of days after his return from one of these mysterious journeys. He rarely traveled about, or visited his neighbors. He lived by levying contributions on the people, or by voluntary presents. His only beverage was chocolate, and the cacao was contributed as voluntary gifts from people far and near. If he entered a house, and offered to buy, or expressed even admiration for anything, whether a chicken, a pig, or any other object, it was at once presented to him. It was considered as good as forfeited. If not presented, it would be sure to die anyhow, and his ill-will would be gained besides. In case of any public calamity, like an epidemic disease, or a scarcity of food from drought, the chief only must visit him, and beg his intercessions with the spirits. He would pay no attention to private appeals. In case he felt inclined to be gracious, he would retire to his cave, and in due time after order a fast. The young man who now holds the position, is one of the finest looking men in the country. He is tall and well formed, his good-natured looking face bears an expression of seriousness hardly in keeping with his youth ; and his whole bearing is grave and impressive. I was forcibly struck by his manner, being so strongly in contrast with the light-hearted, talkative character of most of the people. When in

Cabecar he visited us twice, and on neither occasion did he speak, except when spoken to, unless it was to make some remark, in very few words, and in a low tone of voice, to some of his attendants. His dress consisted of a white shirt, not over clean, a woven cotton breech-cloth, a bright-red handkerchief, tied in a roll around his head, and a magnificent necklace of four strands of large tiger's teeth. He sold me two of the strings for half-a-dollar, and I presented him with some trifles, among which was the rather suggestive article, a bar of soap. He accepted them without any acknowledgment. But then they don't know how to say, "thank you."

Next in importance are the *Tsu'-gurs*. These are the ordinary priests, and their duties are confined to officiating at the feast for the dead. Like the preceding, they are hereditary; only members of one or two families can become priests, and these seem to have all descended from a common ancestor. I have already described the performances of the *Tsu'-gur* at the death-feast of Santiago, and there is nothing to add in that connection. Other feasts only differ in the less degree of profusion and the shorter time they occupy. But there is one circumstance of which I have said little, and that has always seemed to me mysterious. Unfortunately, from no want of effort on my part, I was not successful in investigating this more thoroughly. The songs of these priests are in a language, dialect, or jargon, whichever it may be called, in great part unintelligible to the uninitiated. Some words used are in the vernacular, but many of the nouns are peculiar. *Si-bu*, or God, has at least twenty names; many natural objects have names peculiar to the priests, and the difference is so great that not only I, with my imperfect knowledge of the language, but Mr. Lyon, who speaks it as well as an Indian, could not understand even the purport of the songs. These songs are taught by rote to the young candidates to the priesthood, and are always rehearsed by the priests apart, before being sung. I made several efforts to obtain a vocabulary, but in each case was defeated, rather by the want of understanding on the part of the priest, than from any unwillingness to impart what they knew. At last I made an agreement with the most intelligent and best informed of them. He was to visit me at a certain time and answer all my questions —for a consideration. But a severe attack of rheumatism prevented his coming and lost me the last chance. I have no theory to offer as to the origin of this singular fact. But two explanations however, seem possible. Either the whole thing is an invention, which I think hardly probable, or the system is an exotic, and the songs are in the original language of the missionary who introduced it. I can hardly express my regret at failing to obtain some clue to so interesting a problem.

Finally come the *Awas*, sorcerers, or doctors. This is an open profession, and since it requires but little preparation, gives certain privileges and standing, and brings occasional emoluments, it is pretty numerously filled. The fellows are an arrant set of quacks, and I do not believe there is a single one who acts in good faith. Nevertheless, the people as a rule believe in them. Some of the more intelligent or more civilized of the Indians, those who have been most in contact with foreigners, take foreign medicines when sick, but they are the exceptions. Their method of purifying an unclean person has already been described under the heads of child-birth and uncleanness. They also claim to bring or drive away rain. To do this, the doctor must have a pipe full of tobacco, or a cigar. He goes into the open air, smokes, blows the smoke in certain directions, calling out in an imperative tone of voice, "Rain, go to—" whatever place he may see fit to designate. Once when prisoners between two swollen rivers, forced to wait for them to fall low enough for us to ford; one of our few means of amusement was to give one of these fellows, in our suite, a pipe full of tobacco, and set him to clearing up the weather. He would go outside of our little hut, and between the puffs of smoke would call out, "Rain, go to Panama," "go to Chiriqui," "go to Cartago," in short, to every remote place of which he happened to know the name. It took him ten days before his efforts were crowned with success, and when ultimately the blue patches did begin to appear in the sky, he had the effrontery to calmly claim it as his doing! They also claim to "blow" a proposed route of travel, so as to drive away snakes and bring good luck on the route. In this case, the *modus operandi* is practically the same as for the weather. But their master efforts are when charming away sickness. To see the process, two of my companions feigned sickness and called in the services of one the doctors. He caused each one to procure a live chicken. Catching the animal by the neck and heels he made passes all over the body of the patient, in every direction. Any small animal will answer. Sloths, opossums, even young alligators are used, and are said to be equally efficacious.

After some minutes of this manipulation, he lighted a pipe and blew tobacco-smoke at them. Having given them numerous injunctions about diet, such as forbidding the use of coffee, tobacco, pepper, and salt for a day or two, he went outside the house, and spent half the night seated under an orange tree, singing a doleful ditty, enlivened at irregular periods by unearthly howls and groans. His fee for all this was, in addition to the two fowls, used in the ceremony, and which was all he would have received from an Indian, sixty cents from one and forty from the other; the fees being graduated by the gravity of the supposed infirmities. These doctors claim that their powers are based on the magic merits of certain charms they carry about with them. These charms are supposed to be calculi, extracted from the viscera of animals. Our friend, who tried to change the weather, possessed three of these. One purported to be from the liver of a sloth, another from the bladder of some other animal, &c. I examined them with a glass, and am convinced that they were mere fragments of little calcareous veins, common in the metamorphic rocks of the country, and which had been ground smooth by friction. My little knowledge of medicine, and a moderately well-supplied medicine-case, enabled me to make numerous cures, and of course I soon acquired the title of *Awa*. When asked by my brother professionals to exhibit my charms, I always gravely produced my little pocket compass, which, by its mysterious movements, never failed to impress them. I never could persuade the boldest to touch it.

Three kinds of fasts are observed. The first is only when ordered by the *U-se'-ka-ra* on great public occasions. This is general and simultaneous over all the country. Sufficient food is prepared beforehand to last for three days, the usual time fixed. During those three days, no fires are lighted; the food is served and eaten in silence; no unnecessary conversation is allowed; the people stay strictly inside their houses, or if they go out during day time, they carefully cover themselves from the light of the sun, believing that exposure to the sun's rays would "turn them black"; no salt or other condiment is used in the food; no chocolate is drunk, and even tobacco is forbidden. The second kind is similar, though less rigid than the first, and is voluntary; the same restrictions are observed with reference to fires and food, but the people may talk and go out, avoiding, however, carefully all chance of contact with *bu-ku-ru'*. The third is still more limited, and is the individual fast already referred to for cleansing from *bu-ku-ru'*.

The feasts are of two classes; the death feast already described, and re-unions for labor. In the latter case; when a person wants to do an extraordinary piece of work, like clearing a piece of forest for a plantation, he provides a suitable quantity of food, and especially of chicha. On the day appointed his neighbors unite early at his house, or at the spot designated, and work industriously until about noon. All then repair to the house, and, after a good round of chicha drinking, food is served, followed by more chicha. After a while dancing begins, and is kept up as long as the chicha holds out. Sometimes the work is continued for two or three days, but always ends early in the day, the afternoon and evening being devoted to eating and especially to drinking.

No labor can be accomplished without liberal allowances of chicha, and the man who is the most profuse in this respect is the best fellow. A man will sometimes undertake to make his own clearing, unassisted, but it is very slow work, and drags on at the rate of two or three hours' work a day, with many days of rest. The trees once cut down, the man will burn off the brush, assisted by his sons, or sons-in-law, if he has any, and then plants his crop; usually corn for making more chicha. After that it has to take care of itself. He goes out occasionally to hunt, fish, or sometimes to bring a bunch of plantains. When the corn is nearly ripe, the boys have to watch it to scare off the parrots and pigs. If there are no boys in the family, then all hands usually go and occupy a little shed in, or on the edge of the cornfield. They feast on the green and ripening corn until it is too hard to boil, and then collect what has been left to ripen.

The labor of the women is to bring plantains and water, and to cook and wash. They are never required to do work in the plantation, unless it be perhaps, to help gather and to help carry home the corn. All the sewing is done by the men, even of the little shirts or jackets worn by the women. In carrying loads, the women rival the men in power and endurance. It is nothing uncommon to see a woman, with a big load on her back, and her year old baby seated on top, with his little legs dangling over the front edge of the load. The little monkeys ride securely there through the bush and dodge the overhanging vines and branches as expertly as could be done by an old horseman. When working for each other the people use their own machetes and axes, as a matter of course; but when hired by a foreigner, they invariably expect to be furnished with tools by their employer.

Domestic industry is at the very lowest ebb. Manufactures can hardly be said to exist. The only articles made, beyond furniture, arms &c., are hammocks, net bags, cotton cloth, and pottery. All of these are coarse and inferior in quality. None of the skill exhibited by the Guatemalan Indians exists here. The hammocks are made of a coarse twine, derived from the leaves of a species of *agave*, and are loosely woven in a frame, with a needle. They are hardly long enough for an ordinary person to lie at length in them with comfort, and are used more for seats than for sleeping. They are swung between the posts of the house, near the door, and at a height of from a foot to a foot and a half from the floor. Everything is carried in net bags. They are made with a needle of bone and "meshed" like our fish nets. Some of them are very fine and they are of all sizes, from three inches to two feet deep. They are suspended by a string made of the same material, usually an inch wide and woven openly, in the same manner as the hammocks. The material of the finer and ordinary bags is the fibre of a species of aloe, or *agave*, much finer than that used for hammocks, and naturally nearly white. It is usually dyed of various colors to suit the fancy of the maker. The colors are obtained from several of the native plants and are very durable. A coarser kind is made of the same fibre as the hammocks. These are made with larger meshes, and are used to carry plantains, corn &c., from the field to the house.

The people of Tiribi procure all their bags from the Bri-bris, and I believe, their hammocks also. The Valientes, living beyond the Tiribis, in the adjoining parts of the District of Chiriqui, make similar bags, but much finer and more elaborately wrought. The colors in the Bri-bri nets are always arranged in simple bands, while the patterns of the Valiente nets are often complicated and exhibit considerable taste.

Belts, breech-cloths, cloths for wrapping the bones of the dead, and women's petticoats are woven of cotton. The cotton is raised with no care beyond planting a few seeds and allowing the plants to take care of themselves. They grow to the height of ten or twelve feet, and almost every house has a few in its vicinity. The yellow flowers, buds, and open bolls are seen all the year round, together on every tree. The women collect the ripe cotton, pick it from the seeds with their fingers and spin it. The loom is a simple frame of four sticks, the two upright ones are planted in the ground ; the other two rudely tied to these. The warp is wrapped around the two horizontal bars and a simple contrivance of threads is arranged to open and reverse it. The thread for the woof wound on slender sticks is then passed through in the usual manner and driven tight by blows of a smooth stick. The process is exceedingly slow and tedious and I have never seen it performed except by the men. The belts are usually two to three inches wide and four or five feet long. Breech-cloths are about four feet long and a litle more then a foot wide. The cloths for the dead and the women's petticoats are wider and a trifle longer. Except the cloths for the dead, which are woven white and afterwards painted, most of this cotton work is ornamented with colors. Besides native vegetable dyes, the people of Bri-bri buy cotton dyed a dirty purple with the blood of the *murex*. This is procured from the people of Terraba on the Pacific. They also now occasionally buy colored threads of foreign production, especially a rich bluish purple, of which they are particularly fond. All of this weaving is with very coarse thread, nearly as thick as the finer twines used by shopkeepers in the United States for tying small packages. The cloth is consequently coarse in texture and rough in appearance, but closely woven and soft to the feel. It makes excellent towels, though rather heavy for that purpose. The largest piece of work of this kind I ever saw, was a blanket large enough to cover a good sized double bed. It was in possession of an old woman who wanted to sell it to me for a cow, and refused ten dollars cash.

The pottery now made is the coarsest and poorest I have ever seen. None of the finely made and elaborately ornamented vessels found in the *huacas* or graves are made at present. The use, for half a century or more, of foreign cast-iron pots and kettles has restricted this industry, and possibly helped to injure the character of the work. But two or three vessels taken by me from Tiribi graves, certainly not less than fifty or sixty years old, are in no respect superior to those made at the present day. Native earthenware is now only used for receptacles for chicha. The jars are large—say from ten to twenty gallons capacity. The form is very simple, the workmanship is rough, the clay is coarse and badly mixed, the burning is almost always imperfect, and they are always without the slightest attempt at ornament. The jars are moulded by hand, the clay being added spirally, and moulded by the fingers and trimmed with a smooth stick, in exactly the same manner as I have seen done by the negro women in Santo Domingo. After a certain amount of drying, they are burnt in the open air, in a fire of sticks heaped over them. Each jar is burnt separately.

Although not given to unnecessary exertion, these people travel occasionally from house to house, and even make journeys to Terraba and Limon. The laziest will gladly walk for two days to a dance. They also occasionally go off into the less frequented regions to collect sarsaparilla, with which to buy whatever of foreign manufacture they may want, like axes, machetes, cotton cloth, &c. They never travel alone; always two or more going in company. This is a very prudent measure, since accidents are liable to happen, like snake-bites, or a bad fall, and a person alone and disabled in these wilds, would be more than apt to die before he would be discovered. The preparations for a trip into the forest are simple, but require time. If there are no plantains to be found in the neighborhood to which they are going, a large supply is collected. They are skinned, boiled, and dried hard in the smoke of a slow fire. This is to diminish the weight. A sufficient supply of corn is ground and made into a paste, either with or without the admixture of ripe plantain, for chicha. This is done up in bundles of about a gallon and a half in bulk, carefully wrapped in large leaves and tied with strips, torn from the footstalk of the plantain leaf. At last, all being ready, every person loaded with all he or she can carry, they start out, the loads done up in as compact a bulk as possible and carried on the back, suspended from the forehead by a strip of *mastate*, or bark cloth. Each person also carries in the hand a staff, four or five feet long, made of some tough wood. For ordinary purposes, the entire trunk of certain slender palm trees is used. This makes a stick about as thick as an ordinary civilized walking stick, but very strong, and sufficiently elastic to yield a little without breaking. The chiefs and a few other persons of consequence, like the priests, usually carry a stick of the red wood described above. This is neither so strong nor so light as the palm stick, but it is a privilege of rank, and is preferred in consequence. If the party is going on a trading trip--while the stronger members carry the load of sarsaparilla or rubber, still there are always some, either women or boys, who carry the inevitable bundles of chicha paste. Even when going from one house to another visiting, or to a dance, the chicha is not forgotten, unless the distance is so short that they are not liable to become thirsty on the road. On arriving at a house, the party enters without a word, and each person seats himself where most convenient, but as near the door as possible. The owner of the house, or in his absence, his wife or the next most responsible person approaches the new arrivals and salutes with, "You have come;" "I have come;" "Are you well?" "I am well, how are you?" "I am well." If a particular friend, or a person of consequence, he is invited to seat himself in a hammock. The people of less importance are allowed to take care of themselves. In a few minutes the women of the house approach with calabashes or vessels made of folded leaves full of chicha. If chocolate is to be had, it is prepared at once, and offered in place of chicha. This is a delicate attention, only shown to friends or persons of consideration. Common folks must be content with chicha. Whether chocolate or chica, it is served at least three times, at very short intervals, and at last, when you cannot swallow any more, the polite thing is to say to the person offering it, "drink it yourself," an advice usually followed, and which stops the supply. If the people are particularly inclined to be hospitable, and are fortunate enough to be well supplied, it is not uncommon for the visitor to be overwhelmed with little presents of food. I have been presented within half an hour, in one house, with five calabashes of chocolate, at least half-a-dozen quarts of chicha, a dozen or more ears of green corn, and a dozen ripe bananas. The little boys, with whom I made friends, fared sumptuously, for it wasn't polite for me to refuse anything.

The houses of the Bri-bris are usually circular, from thirty to fifty feet in diameter, and about the same in height. They are composed of long poles, reaching from the ground to the apex. These rest on a ring of withes or vines, tied in bundles, eight or ten inches thick, and resting on a series of upright crotched posts, set in the ground in a circle about a third smaller than the outer circumference of the house. Above this ring, if the house is large, are one or two more, according to its size, not resting on posts, but tied to the sloping poles. The whole is thickly thatched with palm leaves, and finished at the apex by an old earthen jar, to stop the leaks. There is but one aperture to the house, and this is a large, squarely cut door, left on one side. Over the door there is sometimes made a little shed, to keep the rain out. The interior is always very dark. Sometimes, among the Bri-bris, instead of building the house in a circular form, it is elongated and has a ridge-pole, but the ends are rounded, and the door is in one of the ends.

Formerly the Cabecar houses were built in the same style; but now most of them are mere sheds, sloping to one side only and open at the ends and in front. The most pretentious house I saw in Cabecar was a roof sloping to both sides from a ridge pole to the ground, but open at both ends. The Tiribi houses are simply a roof raised on short posts, sloping both ways from the ridge but open all around below. Mr. Lyon told me that formerly the Tiribis as well as the Cabecars had round houses like the Bri-bris, but that the present style is due only to carelessness. The tribes are dwindling so rapidly that they seem to have lost heart even in so important a thing as building comfortable houses; and are content to put up with any make-shift that will shelter them from the weather. The Bri-bri houses are not only better constructed but are much better furnished than those of their neighbors. Beds are placed around the house in the space between the posts and the sloping sides. These are made by planting in the ground two sticks, forked at the upper ends; cross-sticks are laid on these, the other ends being lashed with vines to the sloping rafters. Over these two horizontal sticks are placed boards made of the outer shell of a species of palm; or wild cane is lashed close together. In front of the beds are slung hammocks, between the posts, or to the ends of horizontal sticks projecting a little beyond them. The fire is placed opposite the door near the back side of the house. It is kept up by placing close together, the ends of three large logs which are pushed up as they burn off. Over the fire is a barbacue or frame, sufficiently high to let people pass under it. On it is placed food to keep it out of the way of the dogs, pigs, chickens, and ants. The smoke of the fire is sufficient protection from the latter. Back of the fire-place are ranged the chicha jars, two or three in number. Being round bottomed, they stand on the floor propped up by stones. Scattered around the house are stools or benches, rarely more than six inches high, each carved out of a solid block of wood. They generally have four feet, though occasionally a small, roughly made one is seen, with but two feet, and which is only kept in upright position when somebody is sitting on it. The pots and kettles about the fire are all of American cast iron, and vary in size from less than a quart to ten gallons capacity. Hanging from the barbacue over the smoke, is generally seen a cocoanut shell or a leaf bundle full of salt. It is kept here because it is the only place where it will remain dry. Suspended from the roof are baskets of from one to three cubic feet capacity. They are usually made of a peculiar, very hard, and very flexible vine. These are the trunks of the people, and in them are kept their clothing and all of their little personal treasures and ornaments. They are also used for storing corn or other seeds, like beans, the basket being then lined with leaves to prevent spilling. The women also use them for carrying water calabashes. These are either gourds or the shells of the fruit of the calabash tree, with a small round hole cut in one end. One other use of the baskets is to carry loads when the net bags are scarce. These nets are also often suspended about the house in the same manner as the baskets. Axes, always of the make of Collins, of Connecticut, and long machetes, either of this or of some inferior make, are to be found in every house. Collins' hardware has gained a permanent reputation among these people, who will give twice as much for a leather handled machete of this brand, as for any other kind. Of other tools, the most noteworthy is a heavy stick sharpened to a chisel edge at one end and beveled on one side. This is used for making holes in planting corn or plantain sprouts, and the edge is used to beat down high grass. It works almost as effectually as a scythe. Hooked sticks for lifting the iron kettles, others cut with short radiating branches at the end, like a five or six pointed star, for stirring chocolate, and paddles for stirring food are always found near the fire. Calabashes and gourds with small holes cut in one end for water bottles, and other calabashes cut in half for drinking cups, are also found in every house. Food usually, and even drink sometimes, are served in leaves, called in Spanish "platanillo," smaller and tougher, but otherwise resembling those of the plantain. These are dexterously folded so as to hold a quart or more of fluid without spilling.

Of arms, besides the inevitable machete and very good double-barreled guns, they possess bows made of a very tough kind of palm wood. They are straight and usually about five feet long. The string is made of the finer kind of *agave* fibre. The arrows are of three kinds. All have a butt two and a half to three feet long, made from the light flower stalk of the wild cane. This is a mass of pith, with a thin hard shell on the outside, giving the requisite stiffness. They are not feathered. The front end, from two to even four feet long, is made of the same wood as the bow. For fish this is sharpened to a point and is barbed on one, two, or even three edges, or is made round. For quadrupeds, the wood is shorter, not barbed, and is tipped with a lance-like head made by laboriously grinding down an old knife blade to the requisite form. For small birds, the head ends in a broad round button, flat on the face. The Tiribis use also a little arrow, ending in a slightly open bunch of small reeds. These are for killing a fish, common in the Tilorio, never more than five or six inches long, and which rests attached to rocks by a sucking surface. The fish is so small that several points are necessary to the arrow, so that if one does not strike another may. No poison is used on the arrows, and, in fact the people seem to know of none. In their quarrels, a stick is used over six feet long, nearly an inch thick and about two inches wide, and made of the same wood as the bows, arrows, and planting-sticks. It is very heavy and is grasped by the fingers and thumbs of both hands in such a manner that they are guarded from a blow. They guard and strike an "over-blow" always holding by both hands. They are going out of use now that the people have discovered the easier, but more dangerous process of litigation. Cracked heads and broken arms give way to damages. For killing small birds the blow-gun is used. This is a tube seven or eight feet long, made by punching and burning the pith out from the heart of a palm trunk, nearly two inches thick. They are made very straight and true inside, and are provided with a double sight on top, made of two glass beads placed half an inch apart: when finished they are covered with some resin or a species of pitch to keep them from cracking or warping. The missiles are clay balls. These, previously prepared are carried in a little net, with them there are two bone implements. One, simply a straight heavy piece of bone used to drive a ball out of the tube by its weight, in case of sticking. The other is similar in appearance, but the end is worked into a round pit with sharp edges, for trimming the balls to the proper size and shape. During the war between the Bri-bris and Tiribis, at the beginning of this century, the principal arm used was an iron-headed lance fastened to a shaft barely four feet long. For defense, round shields were carried on the arm, made of the thickest part of the hide of the tapir. I was fortunate enough to secure specimens of both, together with nearly all the other implements, &c., described in the present paper. They are all in the Smithsonian Museum.

All people have some kind of music which doubtless gives pleasure to them, although to our unappreciative ears it may sound rude and disagreeable. The Marimba, an African instrument, found all over semi-civilized Central America, is unknown here. I cannot understand the surprise of an eminent African traveler, who writes wonderingly of the coincidence, of finding this instrument in use in Africa and among the Indians of Central America. It was introduced with the African slaves and has been retained among their descendants and neighbors. The savage Indians do not possess it. The drum is their greatest favorite. It is from twenty inches to two feet long, cylindrical for half its length, with a diameter of six or seven inches; it then tapers convexly to near the other end and then widens out a little. The pattern is always the same, and the size varies but a few inches. The larger end is tightly covered with the skin from the belly of the iguana lizard. It is glued on by fresh blood, being held in place with string until dry. A cord tied around each end suspends it loosely from the left shoulder, and it is held under the left arm, being beaten with the tips of the fingers of the right hand. It is used principally to accompany and keep time to singing and is an indispensable part of every feast or gathering of whatever kind. To accompany the invigorating music of the drum and help the din, an armadillo skin is sometimes used. This is scraped over the rings with a large hard bean-like seed. It at least helps to add to the noise, if it does not contribute melody. A little flute, about as musical as a penny whistle, is sometimes added to the concert, though it seems rather to be looked upon as a toy. These flutes are made of a bone of some bird, perhaps a pelican. The bone has half-a-dozen holes drilled in it, and the end is plugged with wax, so as to direct the air to the larger aperture near the end. I bought one from a Tiribi made of a deer's bone. The priests use in their songs a rattle, made of a small pear-shaped tree calabash, lashed to a bone at the small end. This contains a few seeds of the "shot plant," or Canna. It is held upright and solemnly shaken in time with the song until the end of the stanza, when, as a signal for the chorus to strike in, it is given a dexterous twirl, throwing the seeds rapidly around inside. On very solemn occasions a curious box is also used. It is about eight inches long by four square on the end. It is carved out hollow, with a long tongue on one face, isolated by a U-shaped slit. A heavy handle is attached to one end, also carved out of the same block. When used, it is simply struck on the above-mentioned tongue with a bone or piece of hard stick. This is only used on the death of a chief. There is but one in the tribe, and no bribe that I could offer sufficed to buy it.

Fashions in dress change even among savages, at least as civilization approaches. Formerly the dress of the men consisted only of a breech-cloth. It was made of *mastate*, or bark cloth, about a foot wide and seven or eight feet long, tapering at one end. The cloth is made by taking the inner bark of either the India rubber or another tree and beating it with a roughened stick over a log. This loosens the fibre, and renders it soft and flexible. It is then carefully washed until all the gummy matter is washed out. After drying, it is rubbed a little and becomes soft and smooth to the feel. To apply the breech-cloth, the wide end is held against the belly, the remainder being passed between the legs; it is then wound around the waist and the point tucked in; the broad end then falls over in front, for about a foot long, like an apron. When cotton cloth is used, it is simply caught up in front and behind under a cotton belt, with a similar apron in front. Sometimes, for warmth, a shirt of mastate was worn; simply a strip with a hole in the middle for the head, and tied under each arm with a piece of string. Now many of the men have discarded the breech-cloth, and wear cotton shirts and pantaloons, buying the stuff from the traders and sewing them themselves. Others, not so far advanced, wear a shirt and a breech-cloth. Formerly the hair was worn as long as it would grow, sometimes rolled up and tied behind in a knot. Some of the conservatives still stick to the old style and follow this custom yet; others of the men wear their hair in two plaits, but the majority cut it to a moderate length, and either confine it by a bright-colored handkerchief tied round the head in a roll, or wear a hat.

The dress of the women originally consisted of a simple petticoat (*bana*) of mastate. Very few now use this material, preferring the softer cotton cloth of the traders. The favorite color is a dark indigo-blue, with figures five or six inches across, in white. The *bana* is a simple strip of cloth wrapped round the hips, with the ends overlapping about six inches in front. It is suspended at the waist by a belt, and reaches more or less to the knees. When on a journey in rainy or muddy weather, I have seen a simple substitute. It was made of a couple of plantain leaves, stripped to a coarse fringe and wound round the waist by the midribs. With nothing above nor below it, it is the nearest approach to a fig leaf one can imagine. Only of late have the women begun to wear anything above the waist, and even now it is considered hardly necessary. Some of the women wear a sort of loose little jacket, or chemise, very low in the neck and short in the sleeves, that barely reaches the waist and only partially conceals the bosom. I have frequently seen a woman, in the habit of wearing one of these, either take it off entirely, or fan herself with it, if warm, in the presence of a number of men, and evidently innocent of improper intentions, and unaware that she was doing anything remarkable. With this scanty dress, I must do these people the justice of saying that they are remarkably modest, both men and women. In a year and a half of life in their country, traveling constantly with a body of them, bathing, fording rivers, living in their houses, and seeing more than strangers generally do of the intimate domestic life of the people they are among, I can only recall a single instance of carelessness, and not one of a wanton exposure of those parts of the person, that their ideas of modesty required to be kept covered.

The dress just described is that of the Bri-bris and Cabecars. The Tiribi men, where they do not wear pantaloons, always use the native cotton breech-cloth, never the mastate. The women wear a long strip of cotton cloth, made with a hole in the middle, like a poncho, and reaching before and behind, nearly to the ground. It is gathered up at the waist by a belt, and the edges are caused to overlap at the same time, so that the whole person is securely covered. I was also told that under this they wear a species of breech-cloth or drawers. They are much more retiring in their manner than their Bri-bri sisters; never speak to a stranger except when spoken to, and then reply in as few words as possible and with apparent bashfulness.

For ornaments, all wear necklaces. The favorite ones are made of teeth, of which those of the tiger are most highly prized. Only the canine teeth are used. Small strings are sometimes made of monkey, coon, or other teeth, but are not much thought of. I have seen one of these made of five strings of tiger teeth, gradually diminishing in size, and covering the entire breast of the wearer. The women rarely, almost never, wear these. If they wear teeth, they are of some very small animal. In place of them, they use great quantities of glass beads. I have seen fully three pounds of beads around the neck of one old woman, and she was the envy of all her friends and neighbors. Even little girls are often so loaded down that the weight must be irksome to them. Money is often worn by the women. On one occasion I paid a man six dollars, all in Costa Rican quarters, for his month's work. After a few days I went to his house and saw the entire sum strung on his wife's neck. Shells are also sometimes, though rarely used. The men sometimes carry, suspended from the necklace, the shell of a small species of *murex*, with the varices ground off and a hole drilled in it to make a whistle. These are bought in Terraba, and are highly prized.

The men sometimes wear head-dresses made of feathers. The most highly prized are the white downy feathers from under the tail of the large eagle. Others are made from chicken feathers, or are worked in rows of blue, red, black, yellow, &c., from the plumage of small birds. I have seen one head-dress made of the long hair from the tail of the great ant-eater, in the place of feathers. The feathers are secured vertically to a tape and extend laterally so as to reach from temple to temple, curling over forward at the top, the tape being tied behind, so as to keep the hair in place.

Painting is somewhat in vogue, to assist in the adornment of the person, but is not confined to either sex. The commonest manner is to color each cheek with a square or parallelogram, about an inch across, either solid or made up of bars. This is done with the dark reddish-brown sap of a certain vine, and the pattern resists wear and tear, and water for a week or more. Anatto is also used, but more rarely, and is applied in bars or stripes to the face, according to the skill or taste of the artist. Besides, a hideous indigo-blue stain from a fruit, is sometimes smeared on the face or body, but even savage taste does not seem to approve of this, since it is very unusual.

Formerly the Tiribis tattooed small patterns on their faces or arms; but the younger people have not kept up the custom, and the Bri-bris and Cabecars say they never did it. The chiefs on great occasions wear gold ornaments, similar to those now found in the *Huacas*, or graves of Chiriqui. Whether these have been recovered from some of these graves, or whether they have been handed down from time immemorial is not known. There are but four or five in the tribe, and two of these belong to the reigning chief. The others were formerly also property of the chiefs, but are said to have been given as rewards of merit to the most successful leaders in the Tiribi war. The two belonging to the chief, as well as one belonging to the descendants of one of those warriors, all represent birds. The people call them eagles. The largest is between three and four inches across; the smaller of the chief's two, is double-headed. In connection with these "eagles" another royal emblem might be mentioned. It is a staff of hard black palm wood, over four feet long. The top is carved in the shape of an animal, not unlike a bear sitting on his haunches. But there are no bears in this country, and it must have been intended for some other animal. Below this figure, the stick is square, and is carved out into four pillars several inches long, with spaces between them. In the interior, between them, is a cavity in which a loose piece of the same wood can be shaken about. It was evidently left there in the carving, after the fashion of the Chinese. Below this, the stick is plain. I tried every means in my power to obtain this, but could not buy it.

Games of chance or of skill are equally unknown, and even when brought into contact with civilization, they do not seem to take kindly to gambling. In fact, they have so little to win or lose, and that little is so easily obtained, that the inducement does not exist.

Their food is simple in material and there is but little variation in the manner of preparation. Of meats, besides chickens, they have beef and pork, which are however rarely used except at feasts. They know nothing of salting meat for future use and can only consume one of these animals when a large number is together. Besides the scarcity of beef is so great that probably no Indian possesses more than one or two animals at a time. Wild meat, like peccary, red monkey, (the other species are rarely eaten,) tapir, tiger, even otter, armadillo, and some other small animals are occasionally shot. In this case, all of the meat that is not eaten at once is dried as hard as a bone, and perfectly black, in the smoke of a slow fire. Larger species of birds like curassow are also treated in the same way. It is an interesting fact, universally attested, that the bones of this bird are absolutely poisonous to dogs, while the meat, though tough, is not unpalatable and is perfectly innoxious to man. After a meal it is the never-failing custom to gather all the bones carefully, and either burn them or place them out of reach of the dogs. I do not know whether the flesh would be equally dangerous, though I doubt if it was ever wasted on a dog. This property is said to be due to some fruit or seed they eat. Of vegetable food, plaintains are the staple. In times of scarcity, bananas take their place, besides being eaten raw when ripe. The Indians also occasionally eat a raw ripe plantain, although they are coarse and the flavor is inferior. The methods of preparation are, roasted green, when they make a poor substitute for bread; roasted ripe, when they are eaten with chocolate, with the idea of sweetening it. They are also boiled green, with meat, with green corn, or even alone. Ripe plantains boiled and mashed, are mixed in equal quantities of corn-meal paste to make chicha, or to bake in cakes. They are also, when ripe, boiled, mashed into a paste, and mixed with water into a gruel. This is drank under the name of *mish'-la*. Maize is raised in considerable quantities, and this really involves four-fifths of all their agricultural labor. The corn is of a variety of colors ; white, yellow, red, purple, blue, and almost perfectly black. Sometimes the ear, rarely more than six or seven inches long, is of a uniform color, but more generally the grains are of two or more colors. It is boiled green and eaten from the cob, and is thus considered a great delicacy. It is, when ripe, ground for all other purposes. The process of grinding is rude and simple in the extreme. If possible, a stone, three feet long and two wide, with a flat upper surface, is procured. In default of this, a broad slab of wood is used. For this purpose, a piece cut from one of the plank-like buttresses of the Ceiba tree is procured, and one side dressed smooth. The remainder of this primitive mill, is a stone, about a foot or fourteen inches long, a few inches less in width and three or four inches thick. One side must be regularly curved. The corn, soaked over night to soften it, is placed on the flat surface and the stone last mentioned is rocked on its edge, from side to side. This is always done by the women. When the corn is sufficiently ground, the paste is put into an iron pot and boiled to mush. If it is intended to make cakes, a part of the raw paste is mixed with an equal quantity of boiled ripe plantain paste, to sweeten it. It is then rolled in plantain leaf and baked in the ashes. When the paste is boiled, sometimes a part of it is separated, thinned to the consistency of gruel, and drunk hot. If it is intended to make chicha for the road, the thick mush is at once mixed with an equal part of ripe plantain paste as before, and tied up in leaves. This will keep sweet for two or three days, but gradually fermentation takes place, and at a week old, it has a not unpleasant sweetish acid taste. When ready for drinking, it is dissolved in cold water to a thin gruel. The taste for it is easily acquired, and I admit, I became very fond of it. It certainly does possess intoxicating properties, but I cannot conceive how any civilized stomach could accommodate a sufficient quantity to produce exhilaration. Still I have seen Indians very happy from its effects. But since I desire these notes to be believed, I do not dare to state the quantity I have seen one of these fellows drink. Were only half the truth told, it would appear incredible. The method of preparing the chicha for use in the house is slightly different. The paste is thinned at once, while yet hot. The plantain paste, also thinned, is poured into the earthen jar with it, and sufficient water is added to bring it to the proper thinness for drinking. To produce rapid fermentation another process is yet necessary, which I saw once at Dipuk on the Uren. A young girl (young girls only, with sound teeth perform this operation,) having previously rinsed her mouth with a little water, sat down on a low stool, with a pile of tender raw corn beside her, and a big calabash in her lap. She chewed, or rather bit the grains from the ear and ejected them from her mouth into the calabash. The rapidity of the process was marvelous. She seemed to shave all the grains from an entire ear almost without stopping. There did not to seem be much chewing done, but of course the object was to obtain the saliva secreted during the operation. As fast as her calabash was full she emptied it into the jar of chicha, and proceeded to refill it. I lay in my hammock fully half an hour watching her until she had finished. The next day that chicha was drank and pronounced excellent. I never tried this kind. Such is the force of prejudice. I learned early to prefer doing my own eating.

Beans are also used to some extent, but the quantity planted is generally small, and the people soon have to return to their regular plantains and chicha. I do not think I ever saw half a bushel of beans together in one house. They are large, dark, and generally mottled. They never become very hard, and are of a very good flavor. Small quantities of sugar cane, of a very excellent quality, are raised, but it is only for the purpose of chewing. They never attempt to make sugar or syrup, although some of the foreigners in their country as well as the negroes on the coast make the latter, and the Indians are perfectly familiar with the process. Of the foreigners in the country, perhaps a dozen in all, sambos or mulattoes, with the exception of Mr. Lyon, all raise rice as one of their most important food-staples. The Indians are fond of it, frequently buy it, but never attempt to cultivate it. Of the less important items, they have the fruit of a species of palm called *du-ko'* (*pejiballe* of the Spaniards). This is a small pear-shaped fruit, growing in great clusters ; it has a thin skin on the outside, and a small round seed in the centre. It may be compared to a diminutive cocoanut, the edible portion corresponding with the fibrous husk of that nut. The seed corresponding with the cocoanut proper, is solid and very hard, but has a pleasant flavor. The fruit is very easily raised, requires no care beyond the first planting, and a little weeding for the first year or two, and yet, except at Sarwe, it is very scarce. It is from the wood of this tree that the bows, the arrow tips, the planting and fighting-sticks, &c., are made. Another species of palm furnishes a food, agreeable to the taste, an excellent salad when properly dressed, a perfect substitute for cabbage when cooked, but withal, as my party discovered on one hard journey we made, not very nutritious. It is the bud of tender, half-formed leaves at the top, and can only be obtained by cutting down the tree. It is similar to the deservedly famous palm cabbage of the West Indies, and differs principally in being only about half as large. We found, after living on it almost alone, for nearly a week, that it was good principally for deceiving one's self into starving on a full stomach. *Kiliti*, or "greens" is a favorite dish, probably not much more nutritious than the last. It is made from various tender leaves, put into a pot with little or no water, and gradually steamed into a paste with their own juice. This is eaten with salt when they have it ; otherwise, without.

Cacao is in great demand. The delicious sub-acid pulp is first sucked from the beans, which are roasted and ground on the chicha board, or stone into a coarse paste. It is the greatest luxury they possess. And still, I have never seen a young cacao tree belonging to an Indian. They depend for their supply on the old trees, planted by past generations. I have known an Indian make a two days' journey to collect a little cacao, when less labor would plant him fifty trees near his house.

Fishing is rarely performed with hook and line. They have two methods. One is to shoot the fish from a canoe (all the canoes belong to foreigners), or from the shore, or a rock. They use very long arrows, described previously, and are quite expert. Another method is to select a channel of the river beside an island. A frame-work is built at each end, of sticks and cane, which extend completely across the stream. When everything is ready, the people stationed at the upper end rapidly cover the frame-work with the leaves of the cane, so as to stop the water running through. Those at the lower frame, also spread on cane leaves, but thinner, only so as to keep the fish from passing through. Both parties must work at the same time, and as rapidly as possible, because as soon as the fish find the level of the water lowering they attempt to escape, and I was told that it has sometimes happened that every fish has gotten away before the dams were finished. In the course of a few hours the water is so low that the fish congregate in the deeper pools and are shot with arrows, or even taken out by hand.

The only divisions of time known are the natural astronomical ones: the day, the lunar month, and the year. A glance at the vocabulary will show that special words are used for day in the abstract as distinguished from night, and for to-day, to-morrow, day after to-morrow, &c., and for yesterday, &c. The month is called by the same name as the moon, "*si.*" The year is counted from dry season to dry season, and is recognized by the ripening of the flower-stalks of the wild cane, on which they depend for arrow-shafts. It is called *da-was'* from this connection.

The local diseases of the country are fevers, acquired by going to the coast; or by the hill people, by going down to the low lands. They sometimes seem to become epidemic, due to an unusually wet season, or to the continuance of the rains throughout what should be a dry season. The summer of 1874 was particularly fatal in this respect. Rheumatism is common, especially with the older men. It is brought on by much exposure to rain, and by wading rivers when heated, on journeys. But the commonest infirmities are indolent ulcers, usually on the legs. They originate from any little scratch or bruise, and are the result of the low vital state of the system, due to a bulky but innutritious diet. A wound which, in a person in good health, would heal in a week, may result with one of these people in a sore lasting years, and perhaps at times involving an area twice as large as the hand.

Of remedies, they may be safely said to have none. They are learning to apply to the traders for medicines for fever. All go to Mr. Lyon in case of snake-bite, and when taken in time, he says he has never failed to cure a case with either ammonia or iodine, as seemed to be indicated. It may be interesting to note that after obtaining no relief with one of these medicines, he has given the other, and with immediate good results. He gives the iodine in the form of alcoholic tincture in 10 to 15-drop doses, every 10 to 15 minutes. Some of them seem to believe in the incantations of the *Awas* or doctors, but foreign medicines are gradually gaining ground over sorcery. For rheumatic pains, headaches, &c., there are two remedies used. The simplest is counter-irritation by whipping with nettle leaves. The other is bleeding. The lancet is made usually from the tongue of a jew's-harp, broken off at the angle and sharpened to a point. This is set at right angles in a little stick for a handle, and is used by holding it over the affected part and striking it briskly with a finger. They never regularly open a vein and draw off a quantity of blood, but every stroke makes a separate puncture, from which only a few drops exude. At Borubeta I saw a man bled to relieve the aching of fatigue in his arms. He had been scraping *agave* leaves, to extract the fibre for hammocks. He had at least fifty punctures made over his two arms.

The natural products of the country are principally sarsaparilla root and india rubber. The sarsaparilla vine is green, angular, and covered with thorns. It grows very long and climbs over bushes and even trees in the more open parts of the forest. At short distances it is jointed, and if it touches the ground every joint sends out a new set of roots. The leaves are large and acuminately oval and have three longitudinal ribs, the midrib and two parallel ones, half way between the middle and the edge. The fruit is round and grows in a cluster something like grapes. The vine has a tap-root, and besides sends out a large number of horizontal roots near the surface of the ground, and from six to ten feet long. The sarsaparilla hunter first clears away carefully all the bushes and undergrowth with his machete. He then, with a hooked stick, digs into the ground at the base of the vine until he loosens the earth and finds where the best roots are. The tap-root is never disturbed, and it is customary to dig up only half the roots at a time, to avoid killing the vine. Having selected those that look most promising, he places his hand under one or two and gently lifting them, follows their course with his hooked stick, loosening the soil and lifting them out, following them to their ends. They are then cut off, the dirt carefully replaced around the vine, and the roots laid in the sun, or hung up to dry. A vine yields generally from four to nine pounds of green roots. When dry they are tied into cylindrical rolls a foot long and four or five inches thick, weighing about a pound.

India rubber is obtained by scoring the bark of the trees obliquely. Several cuts are placed one above another and in pairs converging downwards; the sap being directed in its flow by a leaf placed at the bottom, which serves as a spout, to direct it into the vessel placed to receive it.

When collected it looks like milk. It is caused to coagulate and turn black by the juice of a species of convolvulus. It is generally made into cakes a little over a foot long, about eight inches wide and an inch thick.

It is with these two articles, and an occasional deer skin, that all the purchases are made from the traders. They buy various kinds of cotton cloth for clothing, colored handkerchiefs, needles, thread, machetes, axes, knives, iron kettles and pots, a few medicines, and powder, shot, and caps. Their intertribal trade is still more limited. The Bri-bris sell net-bags and hammocks to the Tiribis, and formerly made the large cotton blankets, already described, for sale in Terraba. They buy in Terraba cows and dogs, murex-shell whistles, murex-dyed cotton, and beads made by rubbing down a small species of shell of the genus *Conus*. Sometimes both the Bri-bris and Cabecars, but especially the latter, carry sarsaparilla or rubber a hard ten-days' journey to Matina, to exchange it for cacao, of which they might have enough and to spare for the mere trouble of planting it. But Indians are, almost without exception, a lazy, miserable, and unimprovable race.

It is perhaps advisable to state that the whole of the present memoir was written in Costa Rica, and it was not until my return to Philadelphia, that I encountered the elaborate compilation of Bancroft, on "the Native Races of the Pacific States." At the date of the present writing, but three volumes of the promised five have made their appearance. While I regret that the information in that work, on the present field is so meagre, and in some respects so different from my own observations, I have said nothing which I wish either to retract or modify. I state nothing but what I have seen and learned while living among the people whom I describe. At the same time I trust that I may not be accused of a spirit of antagonism, in pointing out some of the more serious errors in the work in question, and which, if not corrected, might seriously mislead future students.

Vol I. Chapter VII. p. 684, *et seq.* is devoted to "the wild tribes of Central America," and the Indians living below Lake Nicaragua, and the San Juan River are here designated as Isthmians; an appropriate name, since the family seems to cover all of Costa Rica and most, if not all of the State of Panama. But the map, facing p. 684 is utterly incorrect in so far, at least, as it professes to give the distribution of the Indians of Costa Rica.

The region of Talamanca described by me, as containing the three tribes of Cabecars, Bri-bris, and Tiribis, and known to the Spaniards under the generic term of Blancos, is here given up to the Valientes, who should be placed to the south and south-east of the Chiriqui lagoon; and the Ramas, who live in Nicaragua, back of the Mosquito coast. The central plateau, in which are situated the cities and towns of Atenas, San Ramon, Alajuela, Heredia, San José, Cartago, &c., in short, that occupied by practically the entire Hispano-American population of Costa Rica, is here given to the Blancos, and on the shores of the Gulf of

Nicoya, where at present no Indians live, are placed Orotiñans and Guetares. Further, no tribes are placed in South-western Costa Rica, where the semi-civilized Terrabas and Brunkas live ; but on p. 748, the author states that "dwelling in the western part of the state are the Terrabas and Changuenas, fierce and barbarous nations, at constant enmity with their neighbors." Now the Terrabas, as well as their neighbors the Brunkas, or as the Spaniards call them, the Borucas, live in one or two little villages, and are under the complete control of missionary priests, both ecclesiastically and municipally, and are rapidly losing their language, as they have their savage customs, and are approaching the civilized condition of the villages of Pacaca, Coa, Quiricot, &c., in Costa Rica, where the Indians speak only Spanish, and have even lost the traditions of their former state. Again, the Changuinas formerly occupied the valley of the Changuina or Changina River, the main branch of the Tilorio, on the Atlantic slope, and are either entirely extinct, or only represented by a handful of individuals, swallowed up by the neighboring Tiribis on one side, and the Valientes on the other.

In the proper place I have noted what can be said of the Guatusos; there is nothing to add, until a responsible observer has the good fortune to penetrate their country, and survive to tell his tale.

On p. 793 of Vol. 3, is a very short vocabulary of "the language of the Talamancas," copied from the publication of Scherzer. This traveler did not visit Talamanca, but from internal evidence I believe the words to have been obtained from some of the half-civilized Cabecars of Tucuriqui or Orosi, little villages not far from Cartago. In evidence of its unreliability, I note two or three of the most glaring errors of the list.

"Man signa-kirinema. Woman signa-arágre."

Here *signa*, clearly a clerical error for *sigua*, means foreigner, and the word given for woman—*sigua erákur* means foreign woman. So, the prefix *sa* and *su* before the names of parts of the body is the personal pronoun—our. *Suhu* is *sa hu* "our house." "I *be-he*," is really *thou*, the error arising from the Indian answering *thou*, when he was asked, "how do you say I," the interlocutor doubtless pointing to himself. Fortunately the vocabulary is very short, but I am sure there are not more than three or four words in it that would be intelligible to a Costa Rican Indian.

Chapter II.

THE LANGUAGES OF SOUTHERN COSTA RICA.

Section I.—THE BRI-BRI LANGUAGE.

In the following notes, I have endeavored to embody such ideas and conclusions as I have arrived at while studying the language and compiling the vocabulary. From the difficulty of obtaining information from ignorant people, and from my own, by no means perfect knowledge of the language, possibly errors may have crept in, but while I do not think any important ones will be found, I do not venture to claim infallible accuracy. For a year I labored to find some rule for conjugation, and was obliged, as it were, to educate my informers up to the point of being able to give me information about a subject they had never thought of, and could see no use for. Not content to accept their statements categorically, I watched carefully the use of the verbs in their inflexions, and by dint of cross-questioning a number of people, and rejecting everything that was contradictory, I think the few verbs I have selected are correctly given. I have had the advantage not only of a year and a half in the country, in daily contact with a fellow-countryman who spoke the language fluently, enabling me thereby to learn it; but for two months, in the meantime, while absent, I had several intelligent Indians with me who understood Spanish, and finally, after returning to civilization, I had with me for eight months a native, with whom I talked habitually in his own language, and from whom I obtained many corrections of the errors that a stranger must necessarily make. This boy became an apt teacher and voluntarily set me right whenever he heard me use an incorrect expression.

Counting the few abstract words which have doubtless escaped me, and all the specific names of animals and plants, and many of the latter are made up of an adjective, or the name of some plant, combined with *wak* (tribe), I do not think the language can contain two thousand words, and perhaps not fifteen hundred. In preparing the vocabulary I have rejected most of these specific names, because there is no corresponding English word, and a complete natural history collection, carefully studied by competent students, would be required, so as to obtain an equivalent. Even then it would have been useless, because the names vary locally as much as similar words do in English.

In compound words, I have in most cases pointed out the roots, and separated the component parts by a + sign. Although so much detail may have been unnecessary, the study was interesting to myself, and some of the curious results may also interest others.

There can be no doubt but that this and its allied dialects, like all unwritten languages, are undergoing great changes. The language spoken in Terraba was formerly, and probably not long ago, the same as that of Tiribi. There are marked differences between the Cabecar of Coen and that of the Estrella or North River, and even local differences in the use of *r*, *l*, and *d*, can be observed between the half of the Bri-bri tribe living on the Uren, and the others scattered over the rest of the country. In different districts "a little," *wi ri-wi'-ri* is also pronounced *bi-ri-bi'-ri* and *wi-di-wi'-di*, and many other words especially those with *r* or *d* before a vowel, vary fully as much. As has been justly observed by Max Müller, laziness often helps this. The present name for rain *kon'-ni* for instance, is clearly derived from *kong'-li*. In fact the proof exists in the form of the word for dust *kong'-mo-li*. But *kon'-ni* is easier to pronounce than *kong'-li*, and has taken its place.

It would be an interesting study to trace out the ideas which have influenced the formation of compound words. In Bri-bri, a hill is *kong'-bĕ-ta*, the point of the country; in Cabecar it is *kong-tsu'*, the breast of the country, from *tsu*, a woman's breast. Again in Bri-bri a sharp knife is said to be *a-ka'-ta*, toothed (that it may bite, or cut), the beak of a bird is called its tooth; and the same root (*kwo*) is used for a finger-nail, a fish-scale, a bird's feathers, the bark of a tree, or the rind of a fruit.

Some few words are used in such varied connections that they warrant special notice. Among these are *wo*, *kong*, *ĭ-tu*, *kin*, &c. *Kong* is a part of nearly all words relating to the earth, the sky, the atmosphere, in short the general surroundings. It means the country, the day, the weather. In composition it forms part of the word for a hill, valley, &c. *Wo* means originally round, either circular or globular. It is also applied to almost all masses or lumps; it further forms a component of words having a reference to entirety or completeness; thus alone, it means the human face, in compounds it forms a part of the names of the sun and moon, of many parts of the human body, of a drop of water, of a knot, of fruits, seeds, &c; and of verbs, such as to make, to close, to open, to extinguish, to tie, &c.; *ĭ-tu'* means originally to chop, but is applied to shooting, striking with intention of wounding (in contradistinction to *i-pu'* to whip). It also forms part of the verbs to shut, to extinguish, to lie (or throw one's self) down, and, in the latter sense is also used for to pour (to *throw* out of a vessel). *Kin* means a region, or district, and is always used in connection with some qualifying word; thus *Lari-kin*, the country or region of Lari; *dĕ-je'-kin* the salt region or sea; *tsong'-kin* the sand region, or beach; but *nyo-ro'-kin* means in or on the road, and *bĕ-ta'-kin* on top (of a hill or mountain). *Ki-cha'* means originally a string; derivatively a vine to tie with is *tsa' ki-cha*, or a string vine. Veins and tendons are called by the same word on account of their resemblance to strings, while the joints of the limbs are called *ki-cha'-wo* or the lump of strings. *Pa* and *pe*, mean people; the former combined with the 3d person, singular, personal pronoun *ye*, makes *ye-pa*, the 3d person, plural. It is also used combined with *wak*, tribe; thus, *Lari-wak*, means the people of Lari; *sa wak-i-pa*, our people; in this case used probably as much for clearness as anything else, since *tsa-wak*, ("*vine-tribe*") means ants! *Ha-wak-i-pa*, your people. *Pe*, used alone means somebody; whose is it? "*pe cha*;" "somebody's," *cha* being the sign of possession.

There are several words which change their form, or which are even substituted by others, according to the sense or connection; thus *u'-te-kin*, sometimes pronounced *hu'-te-kin*, means out or outside of the house or of anything else in all ordinary cases; but for a person to go out of the house is not *mia u'-te-kin* or *mia hu'-te-kin*, but *miâ hu pa'-gl*. This *pa'-gl* is used in no other connection; and the sound occurs nowhere else in the language except as *pagl-chi-ka* (sugar) and *pagl*, the numeral eight with either of which, it is obviously not related. But the numerals illustrate this most markedly. For instance three is *m-nyat*, and as such it is used in counting all *things*; three houses, *hu m-nyat*; but three men are *pe m-nyal* and three days are *kong m-nyar*. *Bit*, how many, becomes *bil*, how many persons, &c. Old, fat, to grow, pregnant, &c., change in a similar manner when applied to animate and inanimate, or to human and lower objects.

It is remarkable that in a language otherwise so poor, at times it should go to the other extreme. In civilized languages, notably in Spanish, there is a great variety of words to express the shades of colors of animals, particularly the horse. These words, originally adjectives, are often used as nouns. But in Bri-bri we have eight nouns to distinguish pigs, six of which are for color; viz.:

white,	*mu-lush'*.
black,	*do-losh'*.
gray,	*bish'*.
red,	*mash* (*a* as in far).
half-white, half-black.	*bĭ-tsus'*.
black, with white face,	*kŭ-jos'*.
with throat appendages,	*bu-lish'*.
short-legged,	*na'-na* (Spanish *enana*, a dwarf).

These words are in every sense nouns only, and are just as correctly the names of the respective animals as the generic term "*coche*." Chickens and dogs have similar distinguishing names, but I have never been able to learn that horned cattle (*vaca*, whether bull or cow,) are so honored. Horses are comparatively unknown. The only representative of the race in the country being Mr. Lyon's old yellow mare, there has never arisen the necessity for the additional tax on their inventive powers. Words expressing physical qualities of matter are as abundant as in more civilized languages, and their use is as strictly limited. Hard, strong, or stiff, is *dĕ-re'-re*. Soft, like a cushion or fresh bread, is *b-jo'-b jo*, while soft like cloth it is *a-ni'-a-ni* or *a-ni'-ni-ĕ*. Weak or fragile, like a string, or a vessel, powerless like a weak person, or tender like meat, are *to'-to* or *to-toi'*. Elastic, like caoutchouc, is *ki-tsung'-ki-tsung*; when like a switch, it is *kras'-kras*. Plastic, like mud or putty, is *i-no'-i-no*. Pasty, like dough, is *i-tu-wo'*. When more fluid, like very wet mud, it is *a-bas'-a-bas*. Viscid, like syrup or honey, is *kŭ-nyo'-kŭ-nyo*; while very fluid, watery, is *di-se-ré-ri*.

Plantains, bananas, maize, and beans must have been in use by the Indians before the arrival of Europeans, since they have specific names for all of them, but all domestic animals have only the names that came with them.

I have found very few words that I can trace clearly to foreign sources. The names of introduced animals, mentioned above, articles of clothing, and foreign utensils make up almost the entire list. We have *ar'-roz*, Spanish arros'; *sombre'no*, Sp. sombrero; *zapato*, pure Spanish; *pana*, English pan, all hollow vessels of thin metal, of whatever form; *cuchara*, Spanish; *bi-wo*, English bead, *wo* native word for anything round; *tigera*, Spanish; *pussy*, English; *chi-chi*, Aztec *techichi*, the edible dog of Mexico (*fide* Belt), a word used all over Spanish America, and adopted by the Bri-bri and adjoining tribes in the Spanish form; *cachimba*, vulgar Spanish; *da-wa'*, probably corrupted from tabaco; *ko-no'*, corrupted from canoe; *vaca*, *caballo*, and *coche*, Spanish. *Alma*, a corpse, bears a suspicious resemblance to the Spanish *alma*, the soul. *Do ko-ro'*, a chicken, seems to be derived from the crow of the cock; *i-e'-na*, is probably not the Spanish *llena*, with which it corresponds in meaning, but is derived from *e'-na*, finished. *Ese*, that, and *es-es* (= Spanish *eso es*) are probably derived from the Spanish.

The enumeration is decimal, and is simple in structure. Few pretend to count beyond ten, and in counting loose objects if the number is considerable, they are set apart in groups of ten; thus forty-six would be four tens and six. In speaking of numbers the fingers come into play. It is as common to see three, four, or more fingers held up, with the remark "so many" as to hear the numeral mentioned. Beyond ten, the toes are called into service, and the surplus over the ten toes is counted on the fingers, held downwards in this case. The word for five, *skang*, is clearly (*u-ra*) *ska*, the fingers. Beyond ten we have "ten more one," &c., but from twenty upwards I found so much confusion of ideas and contradiction that I strongly suspected my informers of politely trying to invent compounds to please me. By careful questioning, and still better, by watching conversations, I found that twenty is "ten two times," &c., after which the form of the "teens" is repeated; so that twenty-one is "ten two times more one," *d'boƀ but juk ki et*. There is no word for one hundred unless we use *d'bob d'bob juk*, which would be legitimate and intelligible, although I confess I never heard it used.

Wa, *ka*, *ke*, and *ta* added as suffixes are equivalent to the English *ed*. Thus *ĭ-da-wo'*, to die; *ĭ-da-wong̃'-wa*, dead; *lin'-a*, crazy; *ya lin'-a-ka*, he is crazed; *pat'ye*, to paint; *pat-yet'-ke*, painted; *su-tat'*, flat; *sut-tat'-ke*, flattened; *boi*, good; *boir'-ke*, healed; *bĕ-ta'*, a point; *bĕ-ta'-ta*, pointed, &c.

Kli used as suffix is equivalent to our *ish;* thus *boi*, good; *boi-kli*, goodish (*i. e.* pretty good or well); *tyng*, large; *tyng'-kli*, largish; *mat'-ke*, red; *mat'-kli*, reddish. *Ung* and *ong*, which in Terraba and Tiribi are almost the universal signs of the active verbs, are represented by the termination *ung* in nearly a dozen Bri-bri verbs, where it has about the same value as English affix *ate*.

Articles and conjunctions do not exist in the language, the other parts of speech being however present.

Nouns have no inflections for gender, number, person, or case. If it is desired to express sex, the word male or female is used; thus my daughter is called *je la e-ra'-kur*, my woman child; a bull is *vaca we'-nyi* or male cow. The only exceptions to this rule are the few words referring to the human race, like man, woman, and some of the family relationships. Beyond this no distinctions of gender occur.

Number is always indicated by a numeral or by such words as much, many, &c. Two or three words occur that may be considered as apparent exceptions. *Di-cha'* means a bone; *di-che'* is bones. *Di-ka'* is thorn and *di-ke'* is thorns, not two or three, but all the thorns on a tree, in a collective sense. *U-ra'-ska* (*u-ra* arm) is a finger, while *u-ra-shkwe'* (? fingers) is the hand. The coincidence in the termination of these isolated plurals, if they can be so called, is worthy of note.

Person is only indicated by the addition of a personal pronoun. The only semblance of inflection for case, is the addition of *cha*, the sign of possession, alike to nouns and pronouns; or of the prepositions, *wa*, *ta* (with), &c., as suffixes, making an ablative.

The personal pronouns are all monosyllables except *ye-pa* (they), a compound of *ye* (third per., sing.) and *pa* people. Although normally of one syllable, they are often used with the termination *re* (except *ye-pa*) for either emphasis or euphony; thus it is equally correct to say *je* or *je'-re*. *Me* (yourself) is used only in connection with a verb, like *me-sku*, move yourself; *me tu is*, lie (yourself) down. The sign of possession, as stated above, is added alike to the pronoun, or to the name or title of a person; *je-cha*, mine. *Ese* (that) is probably derived from the Spanish, and with *i* (literally *what*) does duty for the neuter. Where the nouns in a language are so simple, it is hardly to be expected that the adjectives and adverbs should suffer many changes. *Boi*, good or well, used either as an adjective or adverb, becomes *boi-na*, better, and a sort of superlative is formed by adding very; *boi chukli*. *Tyng*, large, is in an increased degree either *tyng chukli*, very large, or *tyng bru*; *bru* meaning also large but adding emphasis when the two words are combined. To *boi* and *tyng*, *kli* is added as a suffix to qualify the sense, like *ish* in English; *boi-kli*, goodish, pretty good, and *tyng-kli*, largish, or somewhat large.

The short *i* which begins most of the Bri-bri verbs, is not specially the sign of the infinitive, but is almost universally used where the verb is not preceded by another word, and is sometimes used even then for euphony.

There are four well-defined moods: the infinitive, the indicative, the subjunctive, and the imperative. The subjunctive is as simple as in English, being formed from the indicative by *mi-ka-re'* (if) placed at the beginning of the sentence.

Humboldt,* in speaking of the language of Venezuela, says: "The Chayma and Tamanoc verbs have an enormous complication of tenses," and adds that "this multiplicity characterizes the rudest American languages." It certainly does not apply to the Costa Rican family, which is equally remarkable for the simplicity of its inflections. The present tense does duty for the present participle, and the perfect for the perfect participle; besides which we have the past and but a single future. There is no variation for number or person.

The auxiliaries used are not constant. For the imperative, *ju* is sometimes prefixed, and *mia* is often the sign of the future. It is generally a prefix, but in *ī-haw-na*, to fall, it is added to the end of the word. *Etso* (from *etso-si*, to be,) is the sign of the present tense in *pat-yu*, to paint.

The following examples will give a better idea of the conjugations than a lengthy explanation. They were selected from a large number, and have been verified with as much care as the difficulties of the case would admit. I believe they may be safely trusted, inasmuch as they are words that I have heard in constant use for over two years, and not trusting to categorical information, have watched their habitual use in conversation. The first example, *ī-mi′-a*, is the most variable verb in the language. The forms given in each tense are usable interchangeably. It is equally correct to say, "*je mit-ka*," or, "*je mi-at′-ka*," I go. The past *re*, and *ra′-re*, are used everywhere except by a few people on the Coen River, where the more regular form, *mi-a′-na*, is used.

Conjugations.

To go.

Inf. ī-mi′-a.

Ind. Pres., { mi-at′-ka, mit′-ka, } used interchangeably.

Past, { re, ra′re, } from the verb, *ra′-tski;* the forms ordinarily used.
mi-a′-na; used only on the Coen River.

Perf., mi-cho′.

Fut., { mi′-a, affirmative.
(ke) mi′-na, negative (*ke*, not.)

Imperative, ju. When in combination with an object expressed; *be* JU *i tu*, "thou go shoot." This is the almost universal auxiliary sign of the imperative mood.

ju-shka, *ju*, as above; *shka* (*shku*), to walk.

mi′-shka, confined to the first person plural. It means, "let us go," or, "come," and can be used as an auxiliary to almost all the other verbs; *mi-shka du tu*, "let us go birds shoot."

* Trav., vol. i., p. 327, Eng. Ed.

To burn.

Inf.		ĭ-nyor′-ka.
Ind.	Pres.,	ĭ-nyor-ket′-ke.
	Past,	ĭ-nyor-no′-ka.
	Perf.,	ĭ nyor-no′-wa.
	Fut.,	ĭ-nyor-wa′-ne-ka.

To cook.

Inf.		ĭ-lu′.
Ind.	Pres.,	ĭ-luk′.
	Past,	ĭ-li′-na.
	Perf.,	ĭ-let′-ke.
	Fut.,	ĭ-lu′.
Imper.		ĭ-luk′.

To speak.

Inf.		ĭ-šhtu′.
Ind.	Pres.,	ĭ-šhtuk′.
	Past,	ĭ-šhte′.
	Perf.,	ĭ-šhtet′-ke.
	Fut.,	ĭ-šhte′.
Imper.		ĭ-šhtuk′.

To walk.

Inf.		ĭ-shku′.
Ind.	Pres.,	ĭ-shkuk′.
	Past,	ĭ-shke′.
	Perf.,	ĭ-shket′-ke.
	Fut.,	ĭ-shku′.
Imper.		shku′-ta, walk to (come).
		ju′-shka, walk from (go).

To this verb we must add the following irregular forms: *shkut′-ke*, to walk ahead; its derivative, *it-kat′-ke*, has gone ahead, and *mi′-shka*, for which see the note to the first verb, *ĭ-mia*.

To shoot, to chop.

Inf.		ĭ-tu′.
Ind.	Pres.,	ĭ-tuk′.
	Past,	ĭ-te′-na.
	Perf.,	ĭ tet′-ke.
	Fut.,	(mia) ĭ-tu′.
Imper.		(ju) ĭ-tu′.

To paint.

Inf.		pat′-yu.
Ind.	Pres.,	(etso) pat-yuk′; (*etso*, to be).
	Past,	pat-ye′.
	Perf.,	pat-yet′-ke.
	Fut.,	pat-ye′-ke.
Imper.		pat-yuk′.

To eat.

Inf.		ĭ-kŭ-tu′.
Ind.	Pres.,	ĭ-kŭ-tet′-ke.
	Past,	ĭ-kŭ-te′.
	Perf.	ĭ-kŭ-te′-wa.
	Fut.,	ĭ-kŭ-te′.
Imper.		ĭ-kŭ-tuk′.

To start.

Inf.		ĭ-bĕ-te′
Ind.	Pres.,	ĭ-bĕ-te′.
	Past,	ĭ-bĕ-te′.
	Perf.,	ĭ-bĕ-tet′-ke.
	Fut.,	ĭ-bĕ-te′.
Imper.		ĭ-bĕ-ti′-nuk. Only used in a negative sense, "*ke bĕ-ti′-nuk*," do not start (or move); *i. e.*, "keep perfectly quiet."

To roast.

Inf.		ĭ-ku-ke′.
Ind.	Pres.,	ĭ-ku-kuk′.
	Past,	ĭ-ku-tu′-na.
	Perf.,	ĭ-ku-ket′-ke.
	Fut.,	ĭ-ku-ke′.
Imper.		ĭ-ku-kuk′.

To exchange.

Inf.		ĭ-mne′-we.
Ind.	Pres.,	ĭ-mne-wet′-ke.
	Past,	ĭ-mne′-un͡g.
	Fut.,	(mi′-a) mne′-we.
Imper.		ĭ-mne′-un͡g.

To sleep.

Inf.		kĭ-puk.
Ind.	Pres.,	kĭ-pa-wet′-ke.
	Past,	kĭ-pe′.
	Perf.,	kĭ-pug′-wo. kĭ-pet′-ke; third person plural only.
	Fut.,	kĭ-put′-ke.
Imper.		(ju) kĭ-put′-ke.

To lose (inanimate objects).

Inf.		ĭ-cho′-wa.

There are no changes in this verb, except that *mia* is added to the Ind., Fut. There is no Imperative.

To lose (animate objects).

Inf. ĭ-cho-rai′.
Ind. Pres., ĭ-cho-rai′.
Past, ĭ-cho-rai′.
Perf., ĭ-cho-rat′-ke.
Fut., ĭ-cho-ret′-ke.

To listen.

Inf. ĭ-šhtsu′.
Ind. Pres., ĭ-shtsuk′.
Past, ĭ-šhtse′.
Perf., ĭ-šhtset′-ke.
Fut., ĭ-šhtse′.
Imper. ĭ-šhtsuk′.

To count.

Inf. ĭ-shtaung′.
Ind. Pres., ĭ-shtaunk′.
Past, ĭ-shta′-we.
Perf., ĭ-shtaung′.
Fut., (mia) shta′-we.
Imper. ĭ-shtaunk.

To fall.

Inf. ĭ-haw′-na.
Ind. Pres., ĭ-haw′-nuk.
Past, ĭ-haw′-ne.
Perf., ĭ-haw-net′-ke.
Fut., ĭ-haw′-na (mi), (*mia*).

To push.

Inf. pat′-ku.
Ind. Pres., pat′-kuk.
Past, pat′ke.
Perf., pat-ket′-ke.
Fut., pat′-ke.
Imper. pat′-kuk.

To feed.

jĕ-ku′ has the same terminations as pat′-ku.

To want.

Inf. ĭ-ki-a′-na.
Ind. Pres., { ĭ-ki-a′-na. / ĭ-ki-et′-ke, third person only; when "he *wants* you." }
Past, ĭ-ki-e′.
Fut., ĭ-ki-e′.

The place of the accent is strictly determined by the structure and etymology of compound words. In words composed of a noun and an adjective, the accent is placed on the adjective; thus *di + ki-bi'*, large water, *i. e.*, river; *chi-ka + tyng'*, large substance, *i. e.*, stout; *sa-wi' + juk*, cotton substance or raw cotton. This applies equally to the emphasis in a similar phrase like *pĕ how'-ri*, other, or different people. When the word is composed of an adjective or adverb, with a verb, the accent goes with the verb; thus, *ĭ-shung̃ + pu'*, to spread; *ĭ-wo + tu'*, to shut. When composed of a noun and a verb, it follows the same rule; thus, *bĕ-ta + on'-te*, the remainder (*i. e.*, the end stays or remains). When composed of two nouns, one in an adjective sense, the accent is on the qualifying noun, like *mo' + wo*, navel; *du' + hu*, nest or bird-house; *tsu' + di-o*, milk or teat-juice; *tsu' + wo*, a woman's breast; *tsu-wo' + bĕ-ta*, nipple. This rule is almost universal in Bri-bri, and obtains generally in the other languages; the greatest number of exceptions being in Terraba.

In the simplest sentence, the nominative begins, followed by the object, and the verb comes last. When a noun is qualified by an adjective, the adjective follows the noun. In the same way the adverb follows the verb; and the verb closes the sentence, unless it is accompanied by an adverb, or adverbial phrase. In case there are, in addition to the nominative, object, and verb, another noun, governed by a preposition, these latter close the sentence. I strike you; *je be pu, I thou strike.* I strike you hard; *je be pu dĕrere.* The strong man chops the wood well; *wewi dĕrere kar tu boi.* Will you go with me?; *be mia je-ta*, thou go I with. *Ta, wa*, and *weng̃* (see notes on the nouns) are always added as suffixes to the nouns or pronouns which they qualify, and form a sort of ablative case. But where *weng̃* is used in the sense of "where is," it begins the sentence. Whose hat (is this)? *ji sombreno?* Mine; *je'-cha.* How many people are there in your house? *pe bil tsosi be hu-weng̃?* people how many are thy house-where? Where is he? *weng̃ ye 'tso?* where he is? He remained in the middle of the road; *ye onte nyoro shong*, he remained road middle. Give me a chair (or bench), *krŭ-wa' mu'-nya;* chair give me. Give him, *mu'-ye.* Reach me my hat; *je sombreno be ura reska*, my hat thou hand reach. Heat the water; *di ba-ung*, water make hot. The water is hot; *di ba ba-na*, water warm heated (is). Put out the fire; *bowo wo-tu'*, fire extinguish (or close). The fire went out; *bowo ĭ-to'-wa.* Shut the door; *hu šhku wo-tu'*, house door shut. Unfasten the door; *hu šhku wo-jet'-sa.* Open the door; *hu šhku wo-hu'-wa.* Where is my knife? *weng̃ je tabe?* where my knife (*et so*, to be, understood)? Your knife is there; *be tabe tsosi diya*, thy knife is there. Give me my knife; *je tabe munya*, my knife give. My knife is very sharp; *je tabe akata boi*, my knife toothed good. Go shoot a bird, or go shoot birds; *be ju du tu*, thou go bird shoot. What with? *i-wa?* With a gun; *mokkur wa*, gun with. What kind of a gun? *mokkur is?* gun what kind? Our country gun (blow-gun); *sa konska mokkur*, own country gun. There are no balls (the clay balls or pellets); *mokkur wo ke ku*, gun round (things) no more (*are* understood). Why do you not make some? *i kuenke be ke mokkur wo juwo?* why thou not gun round (things) make? There is no clay (or material); *mokkur wochika ke ku,* gun round (things) material no more. Is your gun a good one? *be mokkur boi?* thou (thy) gun good? Does it shoot well? *ĭtu boi?* shoot well (or good)? Good morning; *be shke'na?* thou art awake, or arisen (literally, straightened up). Reply; *je* (I) *shke'na. Be ratski;* thou hast arrived (salutation on a person entering a house). *Je ratski,* I have arrived. How are you? *is be 'tso?* how thou (*et-so'-si*) art? I am well; *je 'tso boi.* Where did you come from? *weng̃ be bete'?* where thou start? Who went with—? *ji re —ta?* who went — with? I did not see; *ke je wai suna,* not I (*wai* idiom) saw. I do not know; *ke je wai uphchen.* This *wai* occurs nowhere except in these two instances. What did you go for? *iub be re?* why thou went? I went to call my people; *je re je wakipa ĭkiu,* I went I (my) people to call. Are they coming? *yepa ratski?* they come (or arrive? No; I think they have gone away; *au; je hĕnbeku ye micho,* No; I think they have gone. Let us go too; *mishka hekepi,* let us go alike. Where is —? *weng̃—?* He has gone ahead; *ye 't-katke,* he has walked ahead (see note on *ĭ-shku,* in conjugation). Put on your clothes; *be sa-wi' i-u,* thou clothing (cotton) put into.

Section II.—Miscellaneous Notes.

Although the tradition exists that the people of Terraba are a comparatively late emigration from the region of the Tiribis, and although the tradition is sustained by the general resemblances of language, and by the fact that the Brunkas (or Borucas), evidently older occupants of the soil, are crowded into a corner like the Celtic tribes of Europe; yet there are marked differences between the idioms spoken in Tiribi and in Terraba. The Dialects of Southern Costa Rica can be divided into three groups: First, the Bri-bri and the Cabecar; second, the Tiribi and Terraba; and lastly, the Brunka. The three divisions possess many roots and even entire words in common, and may well be compared in their resemblances and differences with the Latin languages. The first group is strongly marked by the short *ĭ* before nearly all verbs and by a generally more musical sound; while the second is harsh, in consequence of the frequent repetition of sound of *z*. The Cabecar *ĭ* before the verb is not so persistent as in Bri-bri, but is more strongly pronounced, approaching more nearly the ordinary Latin or Spanish *i*. The terminations *ung* and *ong* are as marked as the sign of the verb, in the second group, as *ĭ* is in the first. The *z* which almost invariably accompanies this termination, is rarely a part of the last syllable, but is usually sounded at the end of the penultimate, unless when abbreviated into *zu* or *zo*.

A gradual process of change is clearly discernible in these languages. As yet the Bri-bri and Tiribi have been but little affected. But the Cabecar of Coen is absorbing many Bri-bri words because the people of the Coen, although they use their local dialect among themselves, all speak Bri-bri also, while the latter, as the conquerors, despise the Cabecars and never attempt to learn their language. The Cabecars of Estrella rarely speak Bri-bri, but nearly all understand it, as well as Spanish and some speak English, and words of both these latter languages are gradually being adopted. The Tiribis are too isolated to acquire many foreign words; but their near relatives the half-civilized people of Terraba as well as the neighbors of these latter, the Borucas, are rapidly acquiring Spanish at the expense of the corresponding words of their own language. In a party of five Borucas, there was not one who could count except in Spanish; and one of my Terraba friends could remember no word for girl, except *muchacha* (Spanish), until I suggested (supported by analogy) the word *wa-re'* (woman), when he remembered that he had heard some of the old people use *wa-wa-re'!* In like manner, he persisted in giving me the Spanish, "*lucero*" for star, besides many other words.

Many roots run through the entire group of languages unchanged, or with changes so trifling that they are not worthy of note. Again sometimes the root varies while the ruling idea is the same. An illustration of this last case is the following: In Bri-bri, to forget is *hĕn-i-cho*; to remember is *ke hĕn-i-cho*, from *ke* not, *hĕn* the liver, and *i-cho* to lose. To think is also *hĕn be-ku* (probably from *be ket-ke*, ready). Liver in Tiribi is *wŏ*, in Terraba *wo*, and in Cabecar *her*; while to think is, in Tiribi *wo tnizung*, in Terraba *woi-du*, and in Cabecar *her-wik*. The acts of thought, memory, &c., have been attributed to the liver, with about as good reason as we yet place the seat of sentiment in the heart.

In Bri-bri, to lie down is *tu is*, to throw down; imperative *me* (yourself) *tu is*. In Terraba *tush ko* (down) is used in the same manner; *fa tush ko*, thou sit down, and *fa bu tush-ko*, lie down (*bu*) long.

Changes of roots are illustrated by the following. In Bri-bri, *kĭ-puk'* is to sleep, and a hammock is *kĭ-pu'*. In Cabecar a bed is *kă-pu'-gru*, in Tiribi and Terraba it is *bu'-kru*; and in Brunka *kap* is to sleep.

In Brunka a ghost is *i-wik*, and a shadow is *ku-wik'*, and a devil or evil spirit is *kag'-bru*. In Bri-bri, a ghost, or spirit of a dead person is *wig'-bru*. In Cabecar, a shadow is *wig'-ra*, while in Tiribi it is *ya'-gro*, and in Bri-bri, *si-ri-u'-gur*, thus connecting the word in Bri-bri for ghost, or departed spirit, with that for shadow by means of the allied idioms, although without the intermediate changes of the root, it would not have been demonstrable.

It is evident that the Cabecar *mog-i'*, straight, and the Bri-bri *maw'-ki*, true, are identical. Although the Bri-bri word *si'-gua*, foreigner, has been replaced in the other languages, by other words, it remains in the Terraba, as a compound, in the name of the banana, *bin-sigua*, evidently "foreign plaintain," from *bing*, a plaintain; because it may have been introduced at a later date than the larger fruit, and when the word *sigua* was yet in current use.

Again, the idea changes, and with it, words from other roots come in, thus: lightning, in Bri-bri is *ara wo'-nyn*, "the thunder flashes;" the

Tiribi *zhgu-ring'* and the Terraba *zhu-ring'*, seem to be specific; but the Cabecar, *kong-wo-hor'-kn* is "the atmosphere burns," while the Brunka *ji'-kra* is simply "fire."

Like the two or three cases of imperfect plural in Bri-bri, already mentioned, the Terraba has a single plural word; or rather only an approach, a sort of transitional form. *Zhgring* is a rib, and *zhgring'-ro*, the ribs in their collective sense, rather as the bony case of the thorax, than as the several bones.

As stated above, the compound words in the vocabulary of Bri-bri are divided by a + sign between the component parts. In the other languages, there are doubtless many that have not been properly separated, because I have not ventured to make theoretical divisions, and have only separated those that were obviously compound. My less perfect acquaintance with them has not warranted me in this step, nor in the probably unnecessary detail of analysis to which I have subjected the language of Bri-bri.

In Terraba the 3d person, singular, pronoun *kwe*, while not varying for gender or number, has three forms which always appear according to a peculiar condition, thus:

he, she, (sitting or lying down)	so'-kwe.
" " (standing)	shou'-kwe.
" " (going)	her-shon-kwe'.

In Brunka, I, thou, he, (or she) and we, (*a-dĕ-bi'*, &c.,) are used with the termination *dĕ-bi'* whenever they occur alone. When combined with other words in a sentence, the first syllable only (*a*, *ba*, *i*, and *ja*) is used. The termination is almost an integral part of the word and must be used when alone. This is the reverse of the termination *re* in Bri-bri, which is rarely used except in a sentence, and then only for euphony or emphasis, and at the option of the speaker.

Chapter III.

VOCABULARY OF THE LANGUAGE OF THE BRI-BRI INDIANS.

[Note.—In this, and in the accompanying vocabularies, the vowels have the same sounds as in Spanish, unless marked with a special sign; *ĕ* is pronounced as in English *met*; *ĭ* as in *pin*; *ŭ* as in *mum*. *J* has the sound as in *John*; *ng* as in *thing*; *ñg* like the French nasal *n*; *šh* like *ch* in the German *ich*; *h* is aspirated as in English. A few words having unusual vowel sounds are noted separately, not to add unnecessary complication of conventional signs; like *si-ai'*, blue and *ku-ku'*, ear.

Compound words are written with a + sign between the component parts. Accent is of great importance, the change in position of the accent sometimes changing the sense of the word entirely like *ĭ-juk'* to drink, *i'-juk* earth, soil.]

to ache	ĭ-dĕ-li'-na	See *pain.*
to adhere	ĭ-ba'-tsa-wa	See *against.*
afraid	sŭ-wa'-na	
afterwards	e'-wa	Not *i'-wa*, interrogative, "what with."
again	i-să-ka'	See *also.*
against	ĭ-bĕ-tsu'-wa	See *to adhere.*
ago	er'-a-pa	Immediately past time.
	en-i-ai',	Hours ago; this morning.
	nyo-nyo'-ni	Very long ago; days, months, years.
to agree	ĭ-shun'-lu	See *to arrange.*
	nyi'+wo-yu	*Nyi*, together.
to aim	ĭ-shun'-sa-u	
air	koñg'+shu-wang	*Koñg*, see country; *shu-wang*, from *si-wang'*, wind.
alike	nyi+ke'	*Nyi*, together.
	nyi+šhtsei'	Exactly alike; *tsei*, much, applied to words, or two people speaking alike.
alive	tse'-ka	See *awake.*
all	seng	
	o-ri-te-ne'	
alligator	to-rok'	
alone	e'-kur	*E* (*et*) one.
	e'mi	Used in the sense of *only.*
alongside	i-yaw'-mik	
already	je-bak'	
also	i-să-ka'	See *again.*
always	shu-ar'-i-a	
angle	bĕ-ta'	A point; the angle of a surface or the corner angle of a solid.
	si-chi'-a	The angle of a prism; see *square.*
angry	o-ru'-na	
ankle	o-ra-bo'	
ant,	tsa'+wak	*Wak*, people, tribe.
ant-eater	u-ri'	*Myrmecophaga jubata.*
	te'+u-ri	*Tamandua 4 dactyla; te*, a forest clearing; from its being often found in such places.

to arise	ĭ-ku′-ku	
arm	u-ra′	
	u-ra+krong′	Upper arm
	u-ra+nya′-we	Fore-arm, *nya′-we*, belly; see *calf of leg*.
to arrange	ĭ-shun′-lu	To arrange, or agree on a question.
	ĭ-mu boi′-kli-na	There is no one word for to arrange *things* in their places; *i-mu*, to put, *boi′-kli*, pretty good; see introductory notes.
to arrive	ra′-tski	
arrow	ka′-but	Of the various forms of arrows in use, each has also a specific name.
ashes	mu-nu′+chi-ka	*Chi-ka*, material.
to ask	ĭ-cha′-ku	From *ĭ-chu*, to say?
aunt	mi′+a-la	*Mi*, mother; *la*, diminutive.
awake	tse′-ka	See *alive*.
to awake	ĭ-shke′-na	*Shke*, straight.
away	ĭ-mi′+bak	*I-mi′-a*, to go; *bak*(*je′-bak*) already; already gone.
axe	o	Also *shoulder-blade*.
back	shung′+wo	
small of back	ju′+wo	
backwards	tsĭuk′-ā	
bad	su-ru′-i	
	su-ru′-na	Used to express disapproval.
bag	tsku′	A native net bag.
bald	chu-i′	
banana	chi-mu′	
bare	sum′-č	See *naked*.
bark	kar+kwo′-lit	*Kar*, tree; *i-kwo′-lit*, skin.
basket	šhku	
bat	da-kur′	
to bathe	a-kwok′	
to be	et-so′-si	In a place; also *to have*.
beach	tsong′+kin	*Tsong*, sand; *kin*, region.
bead	bi′+wo	*Bi*, (?) corrupted from English *bead*; *wo*, round.
beak of bird	du′+ka	*Du*, bird; *ka*, tooth.

bean	a′-tu+wo	
to bear	su′-na	To bear young (human).
	pa′-na	To bear young (inferior animals).
beard	ka′-luk	
beast	du bi bi′-wak	*Bi*, the devil, or anything mysterious; *wak*, tribe. There is no word exactly equivalent to ours for "beast." Each animal (as well as plant), has it specific name, and *du*, properly belonging to birds, is usually applied if the species is unknown; *bi-wak* is only used in a collective sense.
to beat	ĭ-pu′	To strike, to whip.
	ĭ-bu-ra′+ung	To beat, as on a drum.
bed	a-koñg′	
bee	bur	
	bur′+wak	*Wak*, tribe.
before	keng′+we	*We*, where.
behind	diu′+shent	Behind in the abstract; see *in front*.
	bĕ-ta′+ka	At the tail of a line; immediately behind; *bĕ-ta′*, a point.
belly	nya′+we	*Nya*, see *dung; we*, where.
below	is′+kin	*Is*, down; *kin*, region.
belt	ki-pam′+wo	*Ki-pam*, from *ki-par*, waist.
bench	krŭ-wa′	
to bend	ĭ-wo+shki′+ung	Into a ring; *shki*, a circle.
	ĭ-chung′+wa	To bend at an angle without breaking.
	ĭ-ko-kut′+wa	To bend into a curve.
bent	ko-kutk′	
better	boi′+na	*Boi*, good.
between	shu+shong′	See *middle*.
beveled	sho-utk′	Equally applied to a prismatic solid, or to the cutting off the corner of a surface; see sloping.
bird	du	
to bite	ĭ-kwe′-wa.	
bitter	bĭ-chow-bĭ-choi′	

black	do-ro-roi′	Also very dark blue.
blade	i-wa′	
blind	wo-ju+be′-ie	
blood	pe	
to blow	woi-ku′	With the mouth; *ku* the tongue.
	be-tsir′-ke	*Si-wañg be-tsir′-ke*, "the wind blows."
blue	si-ai′	Last syllable prolonged.
	do-ro-roi′	(Black) very dark blue.
blunt	ke+a-ka′+ta	*Ke*, not; *a-ka′*, tooth; not edged.
	ke+bĕ-ta′+ta	*Ke*, not; *bĕ-ta′*, point; not pointed.
body	wak	Also *tribe*, *race*, *people*.
bog	doch′-ka	See mud.
boil	squek	A furuncle.
to boil	ĭ-tu+wo′	
bone	di-cha′	
bones	di-che′	For notes on this plural, see introduction.
border	iu-ku′	
both	et+et	*Et*, one.
bottle	ko-ku′	See calabash.
bow	shkum-me′	
boy	kŭ-be′	
branch of tree	kar′+u-la	*Kar*, tree; *u-la* (*u-ra*) arm.
brave	we′-bra	
bread	i-nya′	See cake.
to break	ĭ-pa-na′-na	Hard things.
	bu-tsa′-na	A string; *tsa*, a string.
breast	be-tsi′	
breast of woman	tsu′+wo	Also teats of lower animals.
breath	si-wañg	Wind.
breech-cloth	ki-par′+wo	*Ki-par*, the waist.
bright	du-ru′-ru-i	
to bring	ĭ-tsunk′	See to carry.
broad	sho	
broom	wush′+kru	
brother	yil	Always preceded by a proper name or a pronoun.
brother-in-law	ar′-ŭ-wa	

bug	——	There is no generic word. Every prominent species has its name, usually consisting of an adjective, combined with *wak*, tribe.
bundle	dli	
to burn	ĭ-nyor′-ka	
to bury	ĭ-bru′	
bush	kar+tsi′-la-la	*Kar*, tree; *tsi-la-la*, little.
"bush dog"	ro′-buk	*Galictis barbata.*
butt	nyuk	See rump.
butterfly	kwa	
to buy	tu-eng′-ke	
cacao	si-ru′	Also chocolate.
cake	i-nya′	
calabash	ko-ku′	Applied to entire calabashes with a small opening, for water bottles.
	kyong	Cut in half for cups.
calf of leg	klu+nya′+we	*Klu*, leg; *nya′-we*, belly.
to call	i-kin′	To summon, to name.
to call out	i-ya′-na-tsu	The accented *a* like *a* in far.
cane	kar	A walking cane, or stick.
	u-ka′+kur	River cane.
	u-pa′+kur	Sugar cane; see *sugar*.
caoutchouc	si-ni′+chi-ka	*Chi-ka*, material.
care (take)	e′-no-e′-no	
	me+haw′-na-mi	*Me*, yourself; see future tense to fall.
cataract	jol	Also a spring.
to catch	i-krung͡	
centipede	ko	*O* very long.
chaff	i-ku′	
to chase	i-tu′+tiung	
to cheat	wo′-ju	
cheek	onk	
chicha	bo-ro′	A light beer made from maize.
chief	bo-ru′	
child	la′-la	(*Tsi′*) *la-la*, little.
chin	a-ka′+tu	*A-ka′*, teeth.
chocolate	si-ru′	
to chop	i-tu′	Also *to shoot*.

clean	me-ne′-ne ji′-ji i-shung+boi	Also *smooth.* *I-shung,* inside; *boi,* good.
to clean	ĭ-pa′+skwo ĭ-tu′+skwo	*I-skwo,* to wash ; the outside of anything. The inside of a vessel.
clearing	te′	A cleared space in a forest.
close	tsi′-net ku-ke′-ni	Near.
to close	ĭ-wo+tu′	
cloth	di-tsi′ sa-wi′	Made from bark. Made from cotton.
clothing	sa-wi′	Cotton.
cloud	mo shi	The generic word for all clouds. A very dark rain cloud.
club	kir′-u	A long stick for fighting.
coal	bo′-wo+ka	*Bo′-wo,* fire.
cold	se se-seng̑′	Only applied to the atmosphere, as *kong̑+se′,* a cold day. Used in all other connections.
comb	kash	
to comb	kash′+kru	*Kru,* to scrape.
to come	i′-na ĭ-shku′	(Imperative) "come here," To walk.
to complete	o-ro′-na	
compressed	su-tat′+ke	*Su-tat′,* flat.
to consider	be-kĕt-se′-ke	
constricted contracted	su-litk′	Applied to a constriction between two larger parts.
to converse	la′-ri-ke	Only used in the sense of a present participle, conversing.
to cook	ĭ-lu′	
coon	tsi	*Nasua.* There are specific names for the two species, formed by adding adjectives. There seems to be no name for *P. lotor,* which is very rare.
corn	i-kwo′	Maize.

corpse	al′-ma	Can this be Spanish, *alma*, soul?
cotton	sa-wi′+juk	*Juk*, material.
to cough (v) cough (s)	to	The resemblance to the Sp. *tos*, a cough, is probably only a coincidence.
country	kong̑ kong̑+ska	*Kong̑* is used in innumerable compounds. Not only is it used in the same manner in all the allied dialects, but in Brunka, it occurs as *kak*, the sun. Nearly all words relating to country, air, day, atmosphere, sky, earth, in short, the general physical surroundings, contain it as an integral part, *Kong̑+ska* is the country inhabited by any people.
cousin	——	Cousins are called "brother" and "sister," even if several degrees removed.
to cover	pa+bĕ-ku′	*Pa*, skin, covering, surface; *ĭ-bĕ-ku*, see to pack; to cover a solid object.
	ĭ-šhku+pa+bĕ-ku′	To cover a vessel to shut a book.
coward	sŭ-wa+na	See afraid.
crab	ju-wi′	
crazy	i-li′-na	*Ye li′-na-ka*, "he is crazed."
crooked	ki-tunk′	
cup	kyong̑	See calabash.
to cut	ĭ-nyu′	Without chopping.
	ĭ-tu′	With chopping.
cylindrical	a-ra-bo′+wa	
damp	mong′-mok	
to dance	klu′+ptu	*Klu*, the foot; *ptu*, the sole.
dark	tset-tsei′	Also any dark color, especially dark brown.

darkness	koñg+tu-i′-na	"The day darkens" (either from clouds or towards night).
daughter	je+la+ra′-kur	*Je*, my, *la* (*la-la*) son; *ĕ-ra′-kur*, woman. For note on *je*, see *son*.
daughter-in-law	jak′+ĕ-ra	See father-in-law. *ĕ-ra*, (*ĕ-ra′-kur*.)
day	nyi′+we	Contradistinguished from night.
	koñg	Used in all other connections; as *koñg-se*, a cold day.
to-day	in′-ya	
to-morrow	bu-le′	
day after to-morrow	bui′+ki	This *ki*, is apparently "more."
3d day future	m-nyar′+ki	*M-nyat*, three.
4th " "	keng′+ki	*Keil*, four.
5th " "	skang′+ki	*Skang*, five.
6th " "	ter′-i+ki	*Terl*, six.
7th " "	ku′-gi+ki	*Ku′-gl*, seven.
8th " "	pai′+ki	*Pa′gl*, eight.
9th " "	koñg+su-ni′-to	*Su-ni′-to*, nine.
10th " "	koñg+d-bob′	*D-bob*, ten.
11th " "	koñg+d-bob+ki+et′	See eleven.
yesterday	chi+ki′	
day before yesterday	bo′+kli	*Bo* (*but*), two.
3d day past	m-nyon′+li	
4th " "	ka′+ri	
5th " "	skan′+i	
dead	i-da-wo′+wa	See *to die*.
debt	mu′+i	See *money*.
deep	i-shu+tyñg′	*I-shung*, inside; *tyñg*, large; large inside.
	(di)+tyñg′	Deep water.
	wo+kŭ-chutk′	Applied to a *deep* vessel, when the mouth is contracted.
	wo+bli	The same, with the mouth not contracted.
deer	su-ri′	Large species.
	su-ri+ma-ru′	Small species; ma-ru′, reddish.

to depart	mi+cho′	Also perfect, indic., of verb, *i-mia* to go.
to descend	ĭ-u′+mi	*ĭ-u*, to put in ; *i-mi′-a*, to go.
devil	bi	Also ghost, or evil spirit.
dew	mo′+wo-li	*Mo*, cloud ; *wo′-li*, drop.
to die	i-da-wo′	
different	han-′ri	
direction	weng͡	See *where*.
dirt	ka′-mu-ni	
disordered	cho+ri′-li-ĕ	
to dissolve	di+a′-na	*Di*, water.
district	kin	See *region*.
	kong͡	See *country*.
to disturb	ting͡′-we	
to dive	tsant′-kuk	
doctor	a-wa′	
done	o-ro′-ni	Applied to a completed business.
	e′-na	"There is no more."
door	hu′+šhku	*Hu*, house.
double	bit+ung′+wa	*Bit* (*but*) two; *ung*, to make.
to double	ĭ-wo+pung͡′	
down	is	In compounds.
	is′+kin	*Kin*, region (used alone).
to drag	ĭ-ku′+mi	*Mi* (*mi′-a*) to go.
dragon-fly	ki-bi′-a	
to dream	kab′+sueng	*Ka-puk′*, to sleep; *sueng*, to see.
to drink	ĭ-juk′	
to drive	ĭ-bĕ+ku′	*I-ku*, see to drag.
drop	wo′-li	
drum	se-bak′	
dry	si	Like wood, fit for burning.
	si′-na	By evaporation, like clothes after washing.
	po-poi′	Wiped dry.
	mong′-mok	In a less degree than the other words ; but more or less applicable in all cases (partially dry *i.e.* damp). The above are the common usages but are not absolute, the various words being sometimes used interchangeably.

dung	nya	See cake.
dust	koñg′+mo-li	*Koñg*, see note to country; *mo*, cloud; *li* is used in two or three connections with objects in, or derived from the atmosphere, like dew, rain, &c.
eagle	sar′+puñg,	*Sar*, red monkey; *puñg*, hawk.
ear	ku-ku′	*U*, like the German *ü*.
early	bu-la′-mi	*Bu-le′*, to-morrow?
earth	i′-juk	(Soil). Not *ĭ-juk′*, to drink.
earthquake	i	English *e*.
to eat	ĭ-kŭ-tu′	This word is never used in the sense of eating a meal; then *jĕ-ku′*, to feed, is always used.
echo	i-o-ro′-te-nu	
eddy	ir-a-me′	
edge	iu-ku′	
egg	du′+ra	*Du*, bird. In place of "bird," the specific name of the animal is generally given; thus: *to-rok′+ra*, alligator egg.
elastic	ki-tsung′-ki-tsung	Like rubber.
	kras′-kras	Like a switch.
elbow	u-ra+ku-ching′+wo	"Knee of the arm."
	u-ra+knyi′+nyuk	"Heel of the arm."
empty	wu′-ji-ka	See naked.
	wa-ke′-ta	
to empty	ĭ-wu′-ji-ka	
	ĭ-wa-ke′-ta	
	ĭ-tu+tsung	To pour out.
end	bĕ-ta′	Point.
ended	e′-na	"It is all gone."
	o-ro′-ni	Applied to affairs.
enough	wed	
enemy	bo′-ruk	
to envelop	ĭ-bĕ-ku′-wa	
equal, equally	nyi′-ke-pi	*Nyi*, together; *he′-ke pi* alike.
equivalent	ske	
erect	shke′-ka	Perpendicular; see straight.

even	nyi + šhke	*Nyi*, together; *šhke*, level; in a straight line.
	tski-tski'-a	Even in a pile.
	d-ra-d-dai'	Both of these words mean equal on the edges in a pile, like bricks in a wall, or the cut leaves of a book.
	nyi'+es	
evening	tson'-ni	Also late.
to exchange	mne'-we	
to expect	ka'-ble	
to extinguish	ĭ-wo+tu'	Also *to shut*.
eye	wo'-bra	
every	o-ri-ten-e'	See *all*.
	seng̑	
face	wo	See *round*.
to faint	si-wang̑+e'-na	*Si-wang̑*, wind; *e'-na*, to finish.
to fall	ĭ-haw'-na	
family	di-jam'	
far	ka-mi'-mi	
fast	bet'-ku	Rapid.
	dĕ-re'-re	Secure, hard.
fat	ki-u'	Fat, grease or oil of any kind.
	yol'-ta	A fat animal.
	chi'-ka+tyng̑	Fat person; see stout.
father	ji	Always used with a personal pronoun or the name of the person; *je ji*, my father; or with an exclamation, *ah ji*, oh father.
father-in-law	jak	
to fear	su-wa'-na	
fear	su-wa'-na	
feast	sa+bŭ-ra'+ung	*Sa*, we. To feast, to dance and to beat drums are ideas so intimately united in the minds of these people, that the same word is generally used indiscriminately for all three.
feather	du'+kwo	*Du*, bird; *kwo*, see scale, skin, nail, &c.

to feed	jĕ-ku′	See *to eat* and *food*.
female	la′-ki	*La* = *ra* in *ĕ-ra′-kur*, woman.
fever	tak	Spleen.
few	et′+ket wa-wa′-ni	*Et*, one. Also *less*.
fierce	bu-kwe′-wa	
to fight	nyi′+pu	*Nyi*, together; *i-pu*, to strike.
to fill	i-u′	Also *to put in*.
to find	ĭ-kwon′-ju	
fine	wis-wis′-i	Like either a thread, or powder.
finger	u-ra′+ska	*U-ra*, arm.
to finish finished	e′-na o-ro′-ni	See ended.
fire	bo′-wo	
fire-fly	ku′wo ka-tu	Specific. The small flies. The large phosphorescent *elater*.
fire-wood	bo′-wo+tak	*Bo′wo*, fire; *tak*, a piece.
fish	ni-ma′	This is at the same time generic, and is the specific name of the best food fish in the country; the other 15 or 16 species bearing other names.
fish-scale	ni-ma′+kwo	*Kwo*, see skin, nail, &c.
flash	wo′-nyu	
flat	su-tat′ šhke	Like a board, table, &c. Like a floor, a tract of country.
flea	ki	
flesh	du′+ra chi-ka′ du+ra′+chi-ka	*Du*, animal; *chi-ka′*, material; often both words are combined, and more often the name of the animal is used with *chi-ka*, thus *vaca chi-ka*, beef.
floor	hu+shiung	*Hu*, house.
flower	ma′-ma	See plaything.
fluid	di+sĕ-re′-re a-bas′-a-bas	Watery. Like thin mud.
fly	si-chu′	
to fly	i-un′+e-mi	*I-mi′-a*, to go.
fog	mo	See *cloud*.

to fold	ĭ-wo+pung͡g′	See *to double*.
folded	chu-no′-wa	
to follow	ĭ-ju′+ki	
food	jĕ-kuk′	
foot	klu	
force	ke′-sin-kwa	
to forget	hĕn+ĭ-cho′	See introductory notes.
forehead	wo′+tsong͡g	
foreigner	si′-gua	
forest	kong͡g′+juk kong͡g+yi′-ka	*Kong͡g*, see country; *juk*, material.
fragile	to′-to	See *tender*, *weak*.
free	ha′-si	
fresh	pang͡g-ri	
friend	ja′-mi	See *family*.
to frighten	sŭ-wa′+ung	*Sŭ-wa-na*, afraid; *ung*, affix, to make.
frog	ko-ru′	
tree-frog	wĕm	
front	ai-u′-shent	In front, see behind.
froth	i-shu-ji′	
fruit	kar+wo	*Kar*, tree; *wo*, round, a lump.
full	chik-li i-o′-na	This is probably not the Spanish *llena*, but *e′-na*, ended; *i. e.*, "no more can be put in."
gall	šhke	
genitals	ke ma-lek′	Female. Male, human; see *penis*.
to get	ĭ-krung͡g	
ghost	bi wig′-bru	See *devil*.
gift	ti-e′	
girdle	ki-pam′+wo	See *belt*.
girl	ta′-ji-ra ă-la-bu′-si	Before puberty. After puberty.
to give	ĭ-mu′	"Give me," *i-mu′-nya*; "give him," *ĭ-mu′-ye*, or *i-mu-ye-ta*. This is the same word as *ĭ-muk*, to put. To give anything to a person is consequently to put it with him, ĭ-*mu*, to put, *ye*, he, *ta*, with.

glad	ish-tsin′-ĕ	
to go	ī-mi′-a ī ju	For notes on this word see introduction, and especially the conjugation.
God	si-bu′	
good	boi	Also clean, pretty. Emphatic *boi′-hi*.
to grab	ī-krung	
grandfather	re-wu′+je-ke	*Je-ke;* see *old*.
grandmother	nu-wi′+je-ke	
to grasp	ī krung͡g	
grass	kong͡g′+chi-ka	*Chi-ka*, material, is here used contrary to the sense explained, (see material) because *kong͡g +juk*, having the same etymological meaning, is applied to forest.
grasshopper	di′-tsik	
gravel	tsong͡g′+wo	*Tsong͡g*, sand.
grease	ki-u′	See *fat*.
green	tsĕ-bat′-tsĕ-ba	See *wet*.
grief	hed-i-a′na	See *sad*, *sorry*,
to grind	ī-woh′	
to grow	de-tyng͡g′+eh ī-tar+an′-o ī-tar+ar′-ke	A plant. A person or animal.
guatuso	shu-ri′	*Dasyprocta cristata*.
gun	mok′-kur	
hair	konsh′-ko* ko+juk	Of the head. Of the body; *juk*, material. See *leaf*.
half	shong′+buts	*Shong*, see *middle*, *between; but*, two.
hammock	kī-pu′	See *to sleep*.
hand	u-ra′+shkwe	See *finger;* also introductory notes.
handle	kut+a′	*Sister; tube kuta*, knife handle; the sister of the blade!
to hang	ki-chat′+ku ī-mo+wo′+ka	By tying, like a hammock; *ki-cha′*, a string. By simply hooking up, without tying; although *i-wo′-mo* is a knot.

hard	dĕ-re′-re	This word has as many significations as its equivalent in English. It applies to substance, strength, rapidity, and difficulty.
to have	et-so′	See *to be.*
hawk	pun͡g	
he	ye	Also *she.*
head	wo′-ki	
to heal	boir′+ke	*Boi*, good.
heap	i-ra-pa′	
to hear	ĭsh-tsu′	
heart	me′+wo	
heat	ba	
to heat	ĭ-ba′+ung	*Ung*, affix, to make.
heavy	nyets	Usually used with *very: oru+nyets.*
heel	klu+knyi′+nyuk	*Klu*, foot; *nyuk*, butt.
here	i′-nya	In this place.
	i-e′+ku	In this direction; see *there.*
to help	chŭ-ki′-a-mu	
high	kon͡g+shke′	*Shke*, perpendicular.
hill	kon͡g′+bĕ-ta	*Bĕ-ta*, a point; the point of the country; also a mountain.
	u′jum	Applied to all hills or peaks not covered with forest.
hilt	kut+a′	See handle.
hip bone	te′+wo	
hip joint	di-che′+wo	*Di-che*, bones.
to hold	ĭ krun͡g	
hole hollow	ĭ-wo′+an	Any hole, whether a perforation or a cavity.
honey	bur′+di-o	*Bur*, bee; *di-o′*, juice.
hook	bi-ko-ru′	
horizontal	ki-pak′	See *to sleep*, and intoductory notes.

hot	ba ba′-ba	But one syllable is used when in combination with another word, as *kong͡ ba*, hot day; when used alone the syllable is repeated.
	ba+shki-ri′-ri	*Shki-ri-ri*, (*tski-ri′-ri*) yellow; this is used in exaggeration, "yellow hot," as we say "red hot," and is often applied to the weather, food, &c.
	pa+li′-na	(*Ba+i-li′-na*) "boiling hot," similarly used when one is perspiring freely.
house	hu	
how	im′-a	
to hum	ĭ-bor+a-ru′	*Bor*, (*bur*) bee?
humming-bird	bĕ-tsung′	
hungry	dĕ-wo-be-li′-na	
to hunt	ĭ-je-bu′-rik	To hunt game.
	ĭ-ju+lu′	*Ju*, auxiliary; to hunt anything lost.
husband	je+wim′	*Je*, my. See note to son.
hush	sŭ-wang͡+bru′-wo	*Sŭ-wang*, wind.
I	je je′-re	*Re* is a sort of emphasis, added occasionally to all the personal pronouns except *ye-pa*.
if	mi-ka-re′	
to ignore	bru	*Bru ji*, "I do not know who."
	eh′ke	Used only alone, as a reply, while *bru* takes its place in a sentence, as above.
iguana	bwah	
immediately	er′+a-pa	In the past.
	sir′+a-pa	In the future.
in	i-shung͡′	
inclined	o-utk′	See sloping, beveled.

inside	wĕsh′+kin hu′+shung͡ i-shung͡	These two words are applied to the inside of a house; while i-shung͡ is restricted to the inside of a vessel, the interior of the body, of a hollow tree, a box or any other comparatively small space.
instead	ske′	
instep	klu+tsing′	
to interpret	ju-ste′+chu	*I-chu*, to say.
intestines	nya′+kĕ-bi	*Nya*, dung; see *belly; kĕ-bi*, snake.
iron	ta-be′	Also knife; anything made of iron; see *pot*.
it	ĕ-hi′	
jar	ung	
jaw	ka′+ju-a	*A ka*, tooth.
to jerk	ĭ-kunt′-sa	
jigger	ki′+la	*Nigua*; *Pulex penetrans*; *ki*, flea; *la* diminutive.
to join	nyi′+wo-ju	*Nyi*, together; see *to make, to sew*.
joint	ki-cha′+wo	*Ki-cha*, a tendon, a string; *wo*, a lump.
juice	di-o′	Any fluid expressed, like whey from curd; milk from the breast, honey, &c.
to keep	ĭ-bru′	
kidney	bak	
to kill	ĭ′-da-wo′-wa	See to die.
kind	boi′+sen wak	*Boi*, good; in disposition. Class: see *tribe*.
knee	kŭ-chi′+wo	
knife	ta-be′ ta-be′+la	See *iron*. *La*, diminutive; a small knife.
to knock	ĭ-pa′+pu ĭ-bu-ra′+ung	*I-pu*, to strike. See *to beat, feast, to dance*.
knot	ĭ-wo′+mo	*Wo*, round; *mo* (*ĭ-mao′*) to tie.
to know	aph-chen′	
lame	mu′-ya	
language	ŭ-šhtu′	

large	ki-bi′	Simply large. When applied to a stream (*di+ki-bi′*,) it means river, "large water."
	tyn͡g	The commonest form; when applied to water it means deep.
	bru′-bru	Oftenest applied to animals and to domestic utensils.
	tyn͡g′+bru	Very large; more emphatic than the preceding forms.
last	bĕ-te+ka	*Bĕ-ta*, point.
late	tson′-ni	See *evening*.
to laugh	ma-nyu′.	
lazy	jĕ-ke′-i-a	
to lead	u-ra′+yu+mi	*U-ra*, arm; *mi* (*ĭ-mi′-a*) to go.
leaf	sig	Of a plantain, or other large leaf used for wrapper, or for a receptacle for food, &c. The Mosquito word *sic*, from the same root, means a banana.
	kar′+ko-juk	Of a tree, in a collective sense; *kar* tree; *ko′-juk* see *hair*. The idea is the same and the distinction is made by *kar*, the name of a person, a pronoun, &c.
	kar′-ku	*Ku*, tongue; a single leaf.
to leave	ĭ-hu′+unt	*Hu*, house.
left hand	u-ra+bŭ-knick′	*U-ra*, hand, (arm).
leg	klu′+kĕ-cha	
to lend	dĕ-pe′-te-ju	
less	wa-wa′-ni	Really *few*; there is no other word.
to let	on′-si	Imperative; *on′-si tso-si*, *tso-si* (*et-so-si*) to be; "let it alone."
to lick	ĭ-ku′+juk	See *to suck*.
to lie	kon′-shu	
to lie down	ĭ-tu+is′	*I-tu*, to throw; *is*, down.

to lift	ĭ-ku′-ku	
light	su-ru′-ru-i	(White), light colored.
	lu	*Kong*+*lu*, daylight; *bor* +*lu* (*bowo*+*lu*) fire light.
	bo bo′-bra	Light in weight.
lightning	a-ra+wo′-nyn	*A-ra*, thunder; *wo-nyn*, flash.
lips	ku′-kwo	*Ku′*, tongue; *kwo* (*i-kwo-lit*) skin.
to listen	ish-tsu′	
little	tsi′-la-la	
	la′-la	Applied to a child.
	la	Diminutive; used with various nouns; *di*+la, rivulet.
a little	wa-wa-ni wi-ri-wi′-ri wi-di-wi′-di bi-ri-bi′-ri	Local pronunciations.
liver	hĕn	
long	bi-tsing′	
to look	ĭ-saung′	
to look for	lu	Always used with the auxiliary *ju*; *ju*+*lu*.
to loose	ip-tsu′	See *to untie*.
to lose	ĭ-cho′	This is rather a verbal root than an independent word; see to remember and forget. In other cases it carries the terminations *wa*, and *rai*; see notes on the conjugations.
lost	cho′+wa cho+rai′	
louse	kung	
lump	wo	See round.
macaw	pa	Green species.
	kŭ-kong′	Red species.
maggot	hu′+nya	
maize	i-kwo′	
to make,	ĭ-ju+wo′	*Ju*, auxiliary; *wo*, complete.
male	we′-nyi	
man	we′-wi	

many	tsei	See much.
how many,	bit	Impersonal.
	bil	Personal.
so many	ish'-ke	
marsh	doch'-ka	See *mud*, *bog*,
material	juk	Any fibrous, or not compact material; as cotton, *su-wi'+juk;* leaves of a tree, or hair of the head *ko+juk.*
	chi-ka'	Any homogeneous substance; as *si-ru'+chi-ka*, cake chocolate; *su-ni'+chi-ka*, deer meat; *si-ni'+chi-ka*, caoutchouc. Only one exception to this rule exists, see note to *grass.*
meadow	sok	
measure	ya-ma-un'-ya	
meat	du'+ra chi-ka'	See note to flesh.
medicine	kŭ-pu'-li	
metal	nu'-kur	Applied derivatively to money. I have heard quicksilver called *nu-kur'+dio,"* metal juice.
midday	di'+bĕ-ta	*Di-wo*, sun; *bĕ-ta*, point, summit.
middle	shu+shong'	*Shu* is used in nearly all words where the width is a component idea; see *wide*, *narrow*, *between*, *inside; shong*, see *half*, *between.* In a combination, *shong̃* only is used; thus *nyo-ro'+shong*, the middle of the road.
midnight	kong̃+shong'+buts	*Kong̃*, see day; *shong'+buts*, half.
milk	tsu'+di-o	*Tsu*, breast; *di-o*, juice.

mine	je' + cha	*Je*, I; *cha*, sign of possession.
mistake	hen + cho'+wa	See to *forget*, *remember*, *think*.
mole	skwe.	Also rat, mouse, &c.
money	nu'-kur	See *metal*.
monkey	sar	*Ateles*.
	wib	*Mycetes palliatus*.
	hyuk	*Cebus hypoleucus*.
month	si	*Si-wo*, moon. In counting, *si+et* one month, &c.
moon	si'+wo	
more	ki	
	ku	
morning	bu-la'-mi	See early; *bu-le'*, to-morrow.
	en-i-ai'	"This morning," already past; see *to-day*, *here*, *now*.
mosquito	shku-ri'	
mother	je+mi'	*Je*, my; see note to *son*.
mother-in-law	wa'-na	
mountain	kon͡g'+bĕ-ta	See *hill*.
mouse	skwe	Also *mole*, *rat*.
mouth	ku	Of an animal.
	nyuk	Of a river; see rump.
to move	ĭ-sku'	
much	tsot-tsei'	Restricted to quantity or number.
	chuk'li o-ru'-i	Although these refer rather to quality than quantity, they can be used in either sense. When combined, as is sometimes the case for emphasis, they become *o-ru-i chuk'-li*. Although both have the meaning of much, or very, each is used, according to custom, with particular words, although with no difference of sense; *o-ru*

English	Bribri	Remarks
		nyets, very heavy; *tyn͡g chukli*, very large; *pe ratski orui*, many people are coming; *pe tsosi tsot-tsei*, there are many people there.
how much	be-kon͡gs′	
mud	doch′-ka	*Chi-ka*, material.
mute	mĕ,	
nail	u-rats′-kwo	*U-ra+ska*, finger; *kwo*, scale, skin, &c.
naked	sum′-ĕ wu′-ji-ka	Both words are used for bare or naked; but the latter ("empty" *q. v.*) is usually applied to naked children who, according to local custom are yet too small to wear clothing.
to name name	ĭ-kye′ kye	Probably both derived, with ĭ-ki-a′-na, to want, from the same root as *ĭ-kiu*, to call. These three verbs run into each other in conjugation.
narrow	shu+tsi′-la-la	*Shu*, see middle; *tsi-la-la*, small. Anything hollow; also a stream.
	bŭ-sutk′	Anything solid.
navel	mo′+wo	*Knot.*
near	tsi′-net	
	ku-ku′-ni	In place or time.
	ket′-ke	In time only.
neck	ki-li′+kĕ-cha	This *kĕ-cha*, does not seem to be connected with *ki-cha′*, a string or tendon. It occurs again in leg.
necklace	na-mu′+ka	Tiger's teeth.
	pu-li′+ki-cha	Made from shell beads; see *shell* and *string*.
	bi′-wo+ki-cha	Made from beads *q. v.* There are other less common names, all taken from the material

needle	kush	
	di-ka′	Thorn.
negro	tset-tse′+wak	*Tset-tse*, dark ; *wak*, race.
nest	du′+hu	*Du*, bird; *hu*, house.
new	pa′-ni	
night	nĕ-nye′-wi	
nipple	tsu+wo′+bĕ-ta	*Tsu-wo*, breast ; *bĕ-ta′*, point.
no	au	Negation.
	ke	Not.
nobody	ke′+ji	*Ke*, not ; *ji*, who.
noise	ha-lar′	
noon	di′+bĕ-ta	See *midday*.
nose	ji′-kut	
not	ke	
	kam	*A* as in father. Used only as follows—"*kam je bowo′ betse′*" (not I fire prepared). "I have not kindled the fire."
nothing	ke′+ku	*Ke*, not ; *ku*, more.
	shun′-tai	Nothing whatever. Only used for "absolutely nothing."
now	i′-ya	See *here*, and *to-day*.
nuchal lump	ku-li′+duk-wo	*Ku-li*, see neck. The enlarged nuchal ligament caused by carrying heavy loads suspended from the forehead.
numerals		
1	et	
2	but	Impersonal.
	bul	Personal.
	bui	Counting days, future.
	bo	Counting days, past.
3	m-nyat′	Impersonal.
	m-nyal′	Personal.
	m-nyar′	Counting days, future.
	m-nyon′	Counting days, past.
4	keil	
	keng	Counting days, future.
	ka	Counting days, past.
5	skang	
	skan	Counting days, past.

6	terl	
	ter′-i	Counting days.
7	ku′gl	
	ku′gu	Counting days, future.
8	pa′-gl	
	pai	Counting days, future.
	pa	Counting days, past.
9	su-ni′-to	
10	d-bob′	
11	d-bob+ki+et′	*Ki*, more ; *et*, one.
12	d-bob+ki+but′	
13	d-bob+ki+m-nyat′	
20	d-bob+but′-juk	*But juk*, twice.
21	d-bob+but′-juk+ki+et	
oil	ki-u′	
old	i-nu′	Old and worn out, or decayed.
	ke′ji-ke	Old person.
on	bĕ-ta′+kin	*Bĕ-ta′* summit ; *kin*, region.
once	et′+ĕ-kur	*Et*, one ; *e′-kur*, alone.
one, at a time	et+ket′-ke	See *only*.
only	e′-mi	See *alone*.
	ket	*Et+ket*, only one; *but+ket*, only two,
open	ha′-si	
to open	ĭ-šhku+ku′-ka	To uncover a vessel, to open a book ; see to cover.
	ĭ-wo′+wa	To open a door ; see to shut.
	ĭ-shuṅg+pu′	To spread, to unfold.
	ĭ-pu	Also to strike, to push.
	ĭ-wo+pu′	They sometimes say *hu šhku pu*, literally, "push the door (open)," but *i-wo-wa* is better.
to oppose	iu-mu′-ka	
other	să-ka′	*Also*, } There is no nearer way of approaching the idea.
	hau′-ri	*Different*, }
	et′+ĕ-kur	*Once*, }
otter	ha-wa′	*Lutra Braziliensis?*

out outside	u′+te+kin	*Kin*, see region. *u* is probably from *hu*, house. The expression (literally outside of the house) is applied to the outside of anything.
over	bĕ-ta′+kin	See *on*.
oyster	shuk′-te	
to pack	ĭ-be-ku′	See *to drive, to envelop, to cover*.
package	dli	
pain	dĕ-li′-na	See *to ache*.
to paint	pat′yu	
palm of hand	u-ra′+ptu	*Ptu*, palm or sole; see *foot*.
pantaloons	klu′+yo	*Klu*, leg; see shirt.
part	ek′-sin-e	
to part	ĭ-bra′+tu	*I-tu*, to cut.
to pass	ĭ-ru′+mi	*Mi* (*ĭ-mi′-a*) to go.
pasty	i-tu-wo′	Like dough or stiff mud; see *viscid* and *fluid*.
to pay	pa-tu-en′-ke	
pebble	ak′+wo	*Ak*, stone; *wo*, round, lump.
peccary	ka′-sir si-ni	*Dicotyles torquatus.* *D. labiatus.*
penis	ma-lek′ kĕ-be′+wo	Human; see *tail* *Kĕ-be′*, snake; applied to all the lower animals.
people	pe wak wak+i-pa	As individuals, As applied to tribe or race. Collective, thus *sa wak-i-pa*, our people; never *sa-wak*, to distinguish from ant (*tsa+wak*).
perhaps	bru	See *to ignore*.
perpendicular	shke′-ka	See straight.
person	ji ke′-ki.	See who; *ke-ji* nobody. Person of consideration, used like *sir*, in English; probably from *ke′-ji-ke* old.
petticoat	ba′-na	The native dress of the women; a cloth tied round the loins and reaching to the knees.

to pick up	ĭ-shtuk	To gather.
	ĭ-ku'ku	To lift.
piece	tak	
pile	ĭ-ra-pa'	A heap.
to pile up	ĭ-ra-pa'+ung	
piled up	ĭ-ra-pa'+na	
to pinch	ĭ-ku-ni-tsu'-wa	
pine apple	a-mu'+wo	
pipe	ca-chim'-ba	A borrowed word found all over Spanish America.
place	ske,	In *place* of; see *equivalent*.
	i-to'	*Place* for a thing.
	koṅg	See *country*; *ima koṅg kye?* "what is this *place* called?"
plain	koṅg+šhke'	*Shke*, flat.
to plait	du-ki',	
to plant,	ĭ-taung'-bo	Seeds.
	ĭ-kyu	Roots.
plantain	kĕ-rub'	
plastic	i-no'-i-no	
to play	ĭ-nuk'	
plaything	ma'-ma	See *flower*.
	nu'-kur	See *metal*, *money*, and to *to play*.
plenty	o-ru'-i	Much, many.
	shkon'-ten-e	
point	bĕ-ta'	Also summit, top, end.
pointed	bĕ-ta'-ta	
polished	u-ris-u-ris'-i	
possession	cha	See note to *mine*.
pot	ta-be'+ung	*Ta-be'*, iron; see *iron* and *jar*.
to pound	ĭ-wo+tu	
to pour	ĭ-tu'+tsung	To pour out.
	ĭ-tu'	To pour in.
	ĭ-u'	
precipice	ak'+tu	*Ak*, stone, rock.
pregnant	nya'+ye	Human; see *belly*.
	bo'-bo-kye	Lower animals.

to prepare	ĭ-be-ket′-ke	See *ready.*
pretty	boi	See *good.*
price	town′+ske	See *to buy; ske,* value, equivalent.
priest	tsu′-gur	*Ish-tsu,* to sing; a singer.
proof	cha′gu	
to prove	ĭ-cha′-gu.	
to pull	ĭ-kung͡.	
to pull out	ĭ-shung͡′+kung͡	To straighten; to spread out.
pulse	si-wang͡′+ki-cha	*Si-wang͡,* wind; *ki-cha,* string.
to push	ĭ-pat′-ku	
to put	ĭ-muk′	See *to give.*
to put into	ĭ-u′	See *to pour.*
quarter	ĭ-ju-wa′	Applied only to the quarters of an animal; for a fourth part of an inanimate object, they only say *tak,* a piece.
quick	bet′-ku	Rapid, sudden, to hurry.
	dĕ-re′-re	Applied to a rapid stream.
	bou′-i	Very quick.
rain	kaw′-ni	This word is now in a transition state. *Kong͡-+li,* the original form (see note on *dust*) is still sometimes, though rarely, used, and is equally understood.
rainbow	kĕ-be′	Snake.
rat	skwe	Also mouse and mole.
ravine	kong͡+be-li′-na	
raw	ha′-ki	
to reach	ĭ-ru′mi	In going to a place.
	ĭ-re′-ska	With the hand; always used with *u-ra* (arm, hand); thus "I cannot reach it" *ke je u-ra re-ska.*
ready	on′-a	
	bĕ-ket′-ke	To prepare.
red	mat′-ki	
	mat′-kli	Reddish.
	ma′-ru	Brownish red.

region	kin	*Kin* has a double meaning. It is used thus, *Lari kin* the region, or district of *Lari; dĕ-je kin*, the salt region (the sea). Besides it signifies on, or in, a place or direction; *is kin*, below; *bĕ-ta kin*, on the point or summit of a hill; *nyo-ro kin* on the road.
to remain	on′-te	
remainder	bĕ-ta+on′-te bĕ-ta+tso′+nya	Bĕ′-ta, see end, point. *Tso(et-so-si)* to have, to be.
to remember	ke+hĕn-i-cho	*Ke*, not; see *to forget*.
to resemble	sun͡g	To see, to look.
to reside	se′-ne-ke	
to rest	he′-ne-ke	
to return	re′me-li	
ribs	chi-ne′	
ribbed	bŭ-chĕ-no′-noi	
right	boi	Good.
right hand	u-ra+bwa′	*U-ra*, arm; *bwa*, right, in sense of direction or side only.
rim	su-su′-i	
rind	i-kwo′-lit	See *skin, bark*.
ring	shkit′-ke	See *shki*, round.
ripe	ri	
to rise	i-ku′-kn	
river	di+ki bi′	*Di*, water; *ki-bi*, large.
rivulet	di+la	*La*, diminutive.
road	nyo-ro′	
to roast	ĭ-ku-ke′	
rock	ak	Stone.
to rock	a-lik-a-lik′-e	As a cradle, or a round-bottomed vessel.
to roll	ĭ-wo-be-tru′	See *to twist, to turn, to shake*.
roof	hu+ku	*Hu*, house.
roots	wi′+nyuk	*Nyuk*, rump, butt.
rope	bus′-kr du′-ki tsa	A twisted, or "laid" rope. A plaited rope. A common, roughly made rope, a bark string, or a vine used in tying; see vine.

rotten	ĕ-nu′-nĕ-wa	See *old.*
rough	a-ten-ĕ ten-e′	
round	shki	Circular.
	wo	Used for anything rounded, like the face, a seed, a lump in the flesh, a rounded hill, the sun, moon, and in the names of various parts of the body.
rump	nyuk	See *butt, roots, river, mouth.*
to run	ĭ-nen-e′	
sacrum	ju′-wo+di-cha	*Ju′-wo,* small of back; *di cha,* bone.
sad	hed-i-a′-na	See *grief, sorry.*
saliva	wi′-ri	
salt	dĕ-je′	
sand	tsong′+chi-ka	See *beach, gravel, material.*
sap	wu′-li	This root, probably derived from some allied dialect, is now adopted into Isthmian Spanish as "uli," "hule," etc., for caoutchouc.
savannah	sok	
to save	ĭ-bru′	
to say	ĭ-chu′	
scab	ĭ pash′+kwo	*Kwo,* scale; not *ĭ-pa+skwo* to wash.
to scare	sŭ-wa′+ung	See *to frighten.*
scattered	tski′-tski	
scorpion	bi-che′	
to scrape	i-a-pa′+si-u	Like to scrape the bark from a stick; to scale a fish is *i-kwo′+si-u.*
	ĭ-kru′	To clean a dirty surface.
to scratch	i-bi′-u	
sea	di+dĕ-je′	*Di,* water; *dĕ-je′,* salt.
	dĕ-je+kin	See *region.*
to search	ĭ-ju+lu′	See *to hunt, to look for.*
to see	sueng	
seed	wo′	See *round.*
to sell	ĭ-me′-rir	

to send	ĭ-pat-ku+mi	*I-pat-ku*, to push ; *ĭ-mi-a* to go.
to sew	ĭ-wo+ju+wo	*Wo*, besides round, means in this and similar connections, whole, together, complete or closed. See *to close ; i-ju-wo*, to make, "to make closed," or "to make together."
shadow	si-ri-u′-gur	
to shake	i-wo+ti′-u	A violent motion like shaking dust out of a cloth.
	ĭ-wong̃+ju	A gentle motion, like leaves in a breeze.
shallow	i-si′	Applied to water ; *di+si* a shallow stream or pond.
	bu-litk′	A shallow vessel, like a pan or dish.
sharp	a-ka′+ta	*A-ka*, tooth, sharp toothed or edged ; like a knife edge.
	bĕ-ta′+ta	Sharp pointed.
to sharpen	a-ka′+ung	
	bĕ-ta′+ung	
she	ye	Also *he*.
shell	jok′se-ro	Flat univalves ; *helix*, *cyclostoma*, *helicina*, etc.
	pu-li′	Long univalves ; *melania*, *bulimus*, *glandina*, etc.
	su-ri′	*Donax*.
	sa-ra′	Large bivalves.
shield	so′gur	
shin	tang̃′+wo	
to shine	du-r′u′-ru-i	
	ĭ-lu′+gur	*Lu*, light ; to shine like a fire, to give light.
shirt	pa′+yo	*Pa*, skin, covering ; see pantaloons.
to shoot	i-tu′	To cut, to chop.

short	hu′-ye hu′-shi-a	This was explained to me by the person holding his hands but a few inches apart; saying this was *hu′-ye*; with his hands about a yard apart he said *hu′-shi a*, while any greater length is *bi-tsing*, long.
shoulder	so′-bri	
shoulder blade	o	See *axe*.
shrimp	so′	
to shut	ĭ-wo+tu′ ĭ-šhku+pa-bĕ-ku′	See to close, to cover, to open.
sick	ki-ri′-na	
side	wo′+su-li u-ra′	Of the body. Right or left hand; *u-ra*, arm.
silence	bi′-nĕ	
similar	he′+kĕ-pi nyi′+ke-pi di-u′-si nyi+šhtsei′	Alike, also, thus. Equal, alike. "Like that." Exactly alike, in speaking
to sing	ĭsb-tsu′	See *priest*.
sister	kut+a′	
sister-in-law	bo′+kut	
skin	i-kwo′-lit pa	Cuticle, bark, scale, nail, feather, &c. Cuticle, surface, or any soft outer envelope.
skull	wo′-ki+dicha	*Wo-ki*, head; *di-cha*, bone
sky	hoñg′+kut-tŭ	See note to *country*.
to sleep	kĭ-puk′	
sleepy	kĭ-pu+wet′-ke	
sloping	o-utk′	See beveled.
sloth	sĕ-noñg′ se′-ri di′-ra	*Cholœpus Hoffmanni.* *Arctopithecus castaniceps.* *Cyclothurus dorsalis.*
slow	en-ai-en-ai′	
small	tsi′-la-la	See *little*.
small of back	tsiñg-wo	
to smell	la	
to smell good	a-mas-a-mas′ m-nas-m-nas′-i	Like flowers and fluids. Like food.
smoke	shkon-o′	

smooth	ji-ji jis-jis	Both syllables equally accented. Not necessarily polished.
	u-ris-u-ris′-i	Polished.
snail	pu-li′ jok′-se-ro	See shell
	ki-pe	Shell-less species.
snake	kĕ-be′	A curious coincidence exists in the fact that in the Island of Santo Domingo, where there are no venomous reptiles, a poisonous plant, retaining its native name, is called by the people *ki-be′*.
to sneeze	chi′-na	
so	i-nyes′	"So, or thus, he says."
	he′-kĕ-pi	Alike, or similar; it is also used in the sense of "do it *so*."
soft	a-ni′-ni-č	Like cloth.
	b-jo′-b-jo	Like a cushion, or soft bread.
soil	i′-juk	Earth; not ĭ-*juk′*, to drink.
sole of foot	klu+ptu	*Klu*, foot; *ptu*, also palm of hand.
solid	me′-ye	
sometimes	mi-kle′	
son	je+la	*Je*, my; *la*, or *la-la*, from *tsi′-la-la* little. Father, mother, son, &c., are always used with either a personal pronoun, or the name of the relative.
son-in-law	na-wa′-ki-ra	
soon	sir′-a-pa	See immediately.
	tsi′-net	Near.
sore	su-me′+wo	Ulcer.
	ki-nung͡g	Proud flesh.
sorry	hed-i-an′-a	See grief, sad.
sour	shku-shku′-i	
to speak	ĭ-šhtu	
spirit	bi wig′-bru	See *ghost;* also introductory notes.

to spit	wu-ri+tu+wo′	See *saliva*.
spleen	tak	See *fever*.
to spoil	i-nu′-nĕ	See *old*, *rotten*.
spotted	kro′-ro	
to spread	ĭ-shung͡+tsu	Loose objects, as grain, cacao, &c.; also to unroll.
	ĭ-shung͡+pu	A cloth, &c.; see to open.
spring	jol	Also a cataract.
sprit	su-re′+wo	
spy	i-tut′-kuk	
square	shki-shki′-a	This word applies equally to a triangular or a polygonal surface, and means rather *angular*. There are no specific names for figures of different numbers of sides, the exact shape being designated by such phrases as "four-sided," &c.
	si-chit′-ki-a	A square prism; like a beam; see angle.
to stab	ĭ-tiung͡′+wa	
to stand	ĭ-mer′-dwo	
star	bek′+wo	
to start	be-te′	
to steal	hog′-bru	
stick	kar	
to stick to	ĭ-ba′-tsa-wa	
sticky	bi-ti-bi-ti′	
to sting	ĭ-tke′-wet	
to stink	o-ru′+ha-ra	*O-ru′-i*, much.
	la+su-ru′-i	*La*, to smell; *su-ru′-i*, bad.
to stir	ĭ-shu+ĭ-krung͡	*Shu*, see middle; *ĭ-krung͡* to grasp, to hold.
stone	ak	
stool	krŭ-wa′	See *bench*.
stop	pa-pa′	Second person, imperative, present. This verb is used in no other mood, tense, or person. In all other cases, *kin′-tsu*, to wait, is used.
stout	chi-ka+tyng͡′	*Chi-ka*, material; *tyng͡* big

straight	shke+we′	
to straighten	ĭ-shung′-lu	
to strike	ĭ-pu	To beat.
	ĭ-tu	To strike with the intention of cutting or wounding; see *to chop*, *to shoot*, &c.
string	ki-cha′	
strong	dĕ-re′-re	
to suck	ĭ-ku′+juk	*Ku*, tongue; *ĭ-juk′*, to drink; also to lick.
sudden	bet′-ku	Quick.
sugar	pa′-gl+chi-ka	See sugar cane; *chi-ka*, material.
summit	bĕ-ta′+kin	*Bĕ-ta*, point; *kin* region; the summit of a hill or road.
to summon	ĭ-ki-u′	To call.
sun	di′+wo	
sure	jo′-na	See *true*.
to swallow	ĭ-mru′+mi	
sweat	pa+li′-na	See *hot*.
to sweep	ĭ-wush′+kru	See *broom* and *to scrape*.
sweet	bro-broi′	
to swim	a-u′-ku-ri	
to swing	ĭ-ung′-ke-a	
tail	ma-lek′	
to take	ĭ-tsu	
	ĭ-tsu′+me	*Me*, yourself (take from me).
	ĭ-ju′+tsu	*Ju*, auxiliary (go and take.
	ĭ-tsunk′	Take it up.
to talk	ĭ-šhtu′	To speak.
tall	tyn͡g′+bru	See large.
tame	hu′+ru	*Hu*, house.
to taugle	ish-chon′-a-ga	
tapir	na-i′	
to taste	ĭ-quash′-tse	
to tear	ĭ-krash′-a-na	Like cloth.
	ĭ-schi′-na-na	To tear open, like splitting a piece of sugar cane with the hands, or tearing open the skin of an orange.
teat	tsu′+wo	

teeth	a-ka′	While other tribes have special names for the molars, the Bri-bris call them *a-ka+di-u′-shent* (back teeth).
temples	wo′+ki+cha	*Wo-ki*, head ; *kĕ-cha*, see leg, neck.
tender	to′-to	See *fragile*, *weak*.
tendon	ki-cha′	*String*.
testicles	kyak	
that	es′-e	Apparently Spanish, *ese*.
that (is it)	es′-es	" " *eso es*.
then	e′-wa et′-to	Also afterwards.
there	di-ya′ di-ya′+e-ku	"In that direction ;" see *here*.
they	ye′+pa	See *he*.
thick	bu-ri′-ri	
thief	hog′-bru+ru	See to steal.
thigh	tu	
thin	si-bu′-bu-i	
to think	hĕn′+bĕ-ku	See *forget*, *remember*, and introductory notes.
this	i′-sa hi	Not *e′-se*, that.
thorn	di-ka′	*A-ka?* tooth. Derivatively applied to a needle.
thorns	di-ke′	Plural ; see introductory notes.
thou	be be′-re	*Re*, see note to I.
thrice	m-nyat+juk	
throat	bi-do′-nya	
to throw	ĭ-hu′-juk ĭ-tu	See *to shoot*, *to pour*, &c.
thumb	u-ra+ska+wong͡g′-wi	See *finger*.
thunder	a-ra′	
thus	he′-kĕ-pi i-nyes′	See *so*.
tick	bur-ir′-i-e	This is one of several specific names for the same insect.
to tickle	se-cho′-ne	
to tie	i-mao′	

tiger	di-ko′-rum	*F. concolor.*
	na-mu′	Generic.
	na-mu+kro′-ro du-re′grub	*F. onca.*
	se-an′-um	*ditto, black var.*
	išh-tsa+na-mu	*F. pardalis.*
time	nyo-nyo′-ne	Past; it means "a long time ago."
	čn-e′-ri-ĕ	Future time, also remote.
tired	shti-ri′-na	
toad	bu-ke′	
tobacco	da-wa′	
toes	klu+rat′-ska	*Klu*, foot; *rat-ska*, see finger.
together	nyi′-ta	See *with.*
	edj′-ka	
	nyi-šhke′	See *even.*
to-morrow	bu-le′	
tongue	ku	
top	bĕ-ta′	See point, end, summit.
top of head	man-e′+bĕ-ta	*Bĕ-ta*, summit.
torch	kirk	
tortoise	kwi	
to touch	i-ku′+wa	
tree	kar	Also stick; see forest, &c.
top of tree	kar+ko′+bĕ-ta	See *tree* and *summit.*
trunk of tree	kar′+ŭ-ku	
tribe	wak	
true	je′-na	In the sense of "yes, that is so."
	maw′-ki	Absolutely; as contradistinguished from false.
truth	maw′-ki	
to turn	ĭ-wo+tru	See to *twist, to roll, to shake.*
ugly	su-ru′-i	See *bad.*
ulcer	su-me′+wo	
uncle	yĕ-nong′	Maternal.
	yĕ-nong+juk	Paternal.
unclean	nya′	Dirty, filthy; see *dung.*
	bu-ku-ru′	In superstition.
under	is′+kin	See *below.*
to understand	išh-tse′-bo	
unlike	hau′-ri	
unripe	ha′-ki	
	pan′+ri	*Ri*, ripe.

to unroll	ĭ-shung͡+tsu	See *to open, to spread.*
to untie	ĭ-wo′+tsu	
until	ĭa-pan′-a	
to unwind	ĭ-shung͡′+tsu	See *to unroll.*
up	shke a-kong͡	See straight.
upon	a-kong͡ bĕ-ta′+kin	See *point, under,* and *summit.*
upper arm	u-ra′+krob	*U-ra*, arm.
upright	shke′+ka	See *perpendicular.*
to use	ĭ-wa′-tu	
valley	kong͡′+bli	
value	ske	See *equivalent.*
vein	ki-cha′	String.
very	o-ru′-i chuk′-li tu-ru′-ru-i	See much. Applied only to *very* hot water.
vertebra	ko′+wo	
vine	tsa′+ki-cha kar′+ki-cha	*Tsa*, any vine or strip of bark that can be used to tie with; *ki-cha*, a string. *Kar*, wood; generally, one that *cannot* be used to tie with.
viscid	kŭ-nyo′-kŭ-nyo	Like syrup or honey.
voice	or′-ke	
to vomit	cho′+li	*I-cho*, to lose.
to wag	ĭ-wo+tsi′-tsi	Like a dog's tail.
waist	ki-par′	
to wait	ĭ-kin′+tsu ĭ-pan′-a	To wait for anything or person. To wait until another time.
to walk	ĭ-shku′	
to want	ĭ-ki-a′-na	See *to call, to name.*
warm	ba	See *hot.*
to wash	ĭ-skwo′	
wasp	bu-kra′	
water	di	
watery	di+se-re′-re	
wax	bur′+nya	*Bur*, bee; *nya*, dung.
we	sa	

weak	to′-to to-toi′	See *tender; fragile.*
well	ble boi	Noun. Adjective and adverb; good.
to weep	ma-iu′	
wet	nu-ne′-ga tsĕ-bat′-tsĕ-ba	The person, as in a rain. See green; applied to inanimate objects.
what	i ed-i′ wes ji	 "What is it," or "what is the matter." "What did you say?" Personal; who.
when	mi′-ka	
where	weñg we′-du	"Where is —— ?" Used in a sentence. Used alone.
whisper	sa′-sa	
whistle	šhka′-kuñg	*Ka* (*a-ka*) the teeth?
white	su-ru′-ru-i	Also light colored.
who	ji	
whole	wan′-yi	Entire.
why	iub i-kuen′-ke in′-u-i	Used alone, or at the beginning of a sentence, *i-kuen′-ke* means "that is the reason," as well as being used interrogatively. Used alone.
wide	shu+tyñg′	See *middle, narrow,* and *large.*
wife	je+ra′-kur	See *woman* and *son.*
wild	ka+nyi′+ru	See tame; *ka* (*kar*) tree (forest); *nyi,* together.
wind	si-wañg′	
wing	i-pik′	
to wipe	ĭ-pa+kru	See *to scrape.*
with	ta wa	Accompanying. By means of; *i-wa?,* "what with?"
woman	ĕ-ra′-kur	
wood	kar	See *tree, stick.*
to work	ka-nĕ′-bruk	
worm	nya+bus′-ĕri nya′+wak	*Lumbricus; nya,* dung.

to wrap	ĭ-bĕ-ku′-wa	
to wring	ĭ-wo+bĕ-tru′	See to roll.
wrinkled	jŭ-ku-nu-jŭ-ku-nu′	
wrist	u-ra+wo′+bak	
year	da-wab′	The year is counted by the dry seasons when the flower stalks of the river cane are ripe and fit to cut for arrow shafts.
yellow	tski-ri′-ri	Bright yellow.
	dŭ-ko′-lum	Brownish yellow.
yes	he tu	Synonymous ; *hĕ* is most commonly used.
yesterday	chi-ki′	
you	ha	
young	pu′-pu	
yourself	me	Only used in compounds; see note on pronouns.

CHAPTER III.

COMPARATIVE VOCABULARY OF THE CABECAR, TIRIBI, AND TERRABA LANGUAGES OF SOUTHERN COSTA RICA.

English.	Cabecar of Estrella River.	Cabecar of Coen River.*	Tiribi.	Terraba.	Brun-ka (called by the Spaniards Boruca).
to ache				ban	sa-o-ra
to adhere				to-mok′	
afraid	su-wa′-na	———	ban-kret′	ko-ban-kret′	kong′-li
afterwards				dun′-i-ha	ne-nyi′
again	i-shu′-ne	ska-shu′	roz-o′-bi	o-ro-roi	e′-je
against				krŭ-boi′	
ahead			zhug′+ung		
to aim				doz′-ung	
air	kong+shan-ka	kong+shung	kok′+zeng	kong͡-mosh′-ko	
alike				lin-ko-yos′-o-sa	do-she-ri′
alive	ksi	———	si	se	
all	be′-na	———	pir′-kru	tue′	o-ge′
alligator	do-rok′	———	ku	———	ku′-u
alone	e′-kra-bu	e′-kra-i	tok′-sa	———	i-le-shi′
alongside	kot-ke′-mi	———	a-sor-go′		
also				no′-ma	mo-reng′-li
angry				yir′-ke	dre-ha′-lan
ankle	wir-in′-a-kru	———	ton′-kwo	ko-gi-o′	
ant	ksa′+wak	———	sun′-gwo	son′-gwo	jak
anteater (large)	ur-i′	———	shu-go′	———	

* To avoid unnecessary repetition, where the same word occurs in the two dialects of Cabecar, or is the same in the Tiribi and Terraba, a ——— only is used.

English.	Cabecar of Estrella River.	Cabecar of Coen River.	Tiribi.	Terraba.	Brun-ka.
anteater (small)	sa′-sa-ra	———			
any	i-hue′-na	———			
to arise				yir′-ke	
arm	u-ra′	———	ork′-wo	or-og′-o-dok	ju-re′
to arrive	sri′-ska	———	o′-tong		
arrow	u-ka′-wŭ	———	su-re′	———	tun-kas′-a
ashes	mnu′-tu	———	prun′-shuk	prun′-sho	brung
to ask	i-cha-gu′	i-cha-ge′	kag-ra-koz′+ung		tek-shik′
aunt	wai′-bu	———	me′-d-bog-wo	tia (Spanish)	
awake				pe-zhem′-e	
axe				o	
back	shĕ-be′	———	kok-so′	———	ching-kwa′
small of back	tsi-wŏ′	———	si-yek′		a-du-ran′
backwards				i-le′-kung	a-che′
bad	se-ru-i′	———	oi′	so-oi′	sa-i-jen′
bag	ha-mĕ′-tu	———	hru	———	
banana	chi-mu′	———	kei+bing	biu+si′-gua	bri′-dwa
bare	wu′-ji-ka	———	kwor	———	but-sa-she′
bark (of tree)	kar′+kwo	———	kar′+kwo	kor′+kwo-ta	krang′+kwas
basket	ku	———	h-nu′	———	ko-kra
bat	d-kur′	———	i-gur′-gwo	i-gur′	kušh-tsi′
to bathe	bo-ku′	———	we	———	
beach				la-go′	
bead	sa-wa-wu′	sa-wi′-wu	o′-bri	o′-bri-kro	
beak of bird	{ i-ka { du+ka	——— ———	sin′-wa+ko-wo	———	ka-sa

beard	kar′-gu	———	zong	———	ošh′-tsi
beast	bi	———	no generic word	no generic word	
bed	kă-pu′-gru	kong	bu′-kru	{ bu-kru′ i-ong′	
bee	bur+wak	———	or	———	
before	mi-ga′-che	———	bom′-gung	———	
behind	bĕ-ta′+ha-mi	———	ir′-ung		a-wah′-i-gi
belly	nya′+wi	———	bo-wo′	———	ku′-a
below				tush′-ko	
belt				kro-ten′-za	suamp′-ka
bench	kru′-wa	———	kruk	kru	te′-kra
better	boe′-si	———	kob′+ko	kob′-i	
between	shas′-ka	shus′-ka	jig′-rai	zig′-ro	
bird	du	———	sin′-wa	sen′-o-wa	du-tsut′
to bite	i-gu-tu′	———	tun-ez′+ung	we′-a	tu′-li
bitter	b-chin′-a	———	kek-teng′		
black	do-ro′-na	———	zi	sok′-si-ĕ	tu-li-nut′
blade				druñg′-wa	
blind				bo-kwoi-zhem′	tu-ĕt′
blood	pi	———		sring	ji-bi′
to blow	wu′-kwa	———	bu-kwaz′+ong	ba-kwoz -ung	i-bu-hu-grau′
blow-gun	mok′-kur	———	mok′-ru		
blue	siu	———	ding-ding′-eh	ding-ding′-ĕh	wa-a-gat′
blunt	kai+hu-riu′-a-wa	———	beg-jim′-ĕ		
body	wa′-ki	wa-ge′	i-do′	i-dob′	
bold	ĭ-han′-ĕ	———			
bone	di-cha′	———	d-bo′-gro	———	det-kra′
both	bor+ki′	———			

English.	Cabecar of Estrella River.	Cabecar of Coen River.	Tiribi.	Terraba.	Brun-ka.
bow	u-ka′+bĕ-ta	———			
boy	ber+je′	har-a′	du-met′+wa	kwa-zir′	jas-rok
branch (of tree)	kar′+u-ra	———	kor′+ko-wo		
to break	i-pan-o′-wa	———	toz′-ung		
breast	her′+bĕ-ta	———	wor-bu′	———	bi-dran′
breast (of woman)	tsu	———	nok′-a	nok′-o	i-ka′
breech-cloth	ki-par′-wo	———	shwong+kin′-go	shwong-king′	
bright	d-ra′-na	kan-yi-na′	kro′-zhing	———	
to bring	tsum′-be-te	———	tek+suz′+ong		
broad				ba-mi-on′-so	
broom				shto-ko′-kra	
brother	ser′-a	ja-mi′	shi		
brother-in-law	shuo-wa	———	bau′	———	
bug	*nya′-hru	———	*guk′-sha		
bundle				pfon-u′	
to bury	i-tu-bi-u′-wa	———	pnoz′-ong	pfron′-zo	
bush	kar+tsi′-lash-tu	———			
bush dog†	rob-gu′	———		o-ru-bu-gu′	
butterfly	kwa	———	kwong′-wo	———	kwaŝh′-ka
to buy				tun-ez′-ung	
cacao	tsi-ru′	———	kau	ko	kao
calabash	d-ka′	———	yi-guk′	———	jun′-kra
cane (sugar)	pash-tu′	———	se-ror′+bo	———	bah+je-ra′

* These words are probably specific. The genius of this group of languages is not to have generic words like bug, beast, palm, &c.; though there are some exceptions, like bird, butterfly, &c.

† For the technical names of these animals, see the Bri-bri vocabulary.

cane (river)	u-ka′-gru	u-kwa′-gru	se-ror′+gro	s-ro′-gro	bah+kra′
caoutchouc				sŭ-ru′	
to carry	tsum′-mi	———	hek+suz′+ong		
cataract				her-pig′-te	
catarrh	to	———	tau′		
to catch				shaz′-ung	
to cheat	i-hag′-bra	i-kon′-ju-wa	yab-goz′+ung	———	
cheek	hog-o-tu′	———	kwog′-do	kwog′-ro	•pshu-gran′
chicha	bro	———	o	———	jĭ-wo′
chief	bo-ru′	———	pu-ru′	———	
child	jo-ba′	jo-ba-ra′	wa	an͡g′-wa	
chin	ska′-jua	———	o-ro′	———	ksha′-di
to chop	i-bi-ku′-hŭ	———		zog′-ung	
clean	ma-nen-ai′-re	man-em-en-e′		krung′-zo	trut′-ka
cloth (cotton)	sa-wa′+ba-gna	sa-wa′	shkwe′+shwong	shkwe′+bor-shwong	
" (bark)	dŭ-tsi′	———	keb′+kwo	———	
cloud	mo	———	pong	———	bok
coal (of fire)	ji-ko′+wa	ji-ko+wo′	yuk+kur′-kwa	kur′-gra	
cold	tse′-na	———	smo	sa-mo′	tsa′
color	ia-wa-kir′		kro-zher′-ba		
comb		i-aĭ-wa-gir′		kun′-chi	kash
come	hi′-ga	———	pa-teg′-ŭn	———	
to complete				ton-oi′	
coon	si-rak′	———		si-luk	
corn (maize)	i-kwo′	———	ep	———	kup
corpse	i-shen-a′-wa	shin′-a-wa	shin′-mo	———	
cotton	su-wi′+ji-ku	———	shkwi	shwke′-sho	

English.	Cabecar of Estrella River.	Cabecar of Coen River.	Tiribi.	Terraba.	Brun-ka.
cough	to	———	tau	———	
to cough	kor′-har	kor′-her	shtenz′+ung		
to count	kansh′-ta-wa	———	kok-shtoz′+ung		
country*	kong	———	kok	kosh′-ko	kak
cousin	ta-ra′	———	shtob		
to cover	i-ki′-tu-wa	———	kim-yoz′-ong	kin-yoz′-ung	
crab	ju-wi′	———	ye-beg′-wo	———	
crazy	ri-na′-ka	———	skwĕ	———	
to cry	mi-o′	———	so-loi′-wi		
to cut	i-tu′	———	zoz′-ung	———	tsa-a-ran′
to dance	b-klu+tu′	———	zron	———	kwi-gri′
dark	stu-i′-na	———			
daughter	ba′+bu-si	———	wa′+wa-re	———	
day	kan-yi-na′	kan-yi-na′-wa	d-ba′	da-boi′	kak′+a-ba-ra
to-day	hir	———	er′-i	———	cha
to-morrow	bu-ri′-ri	———	i-bong′	i-bon′-a-ga	sek
day (after to-morrow)	boi-ki′	———	kun-ma′	kur+ma′	bwek
third day future				mi-au+ma′	mang-ek′
fourth day future				kin+ma′	
fifth day future				sẖkin+ma′	
yesterday	jĕ-ki	———	kub′-kĕ	kub-kesh′-ko	bi-ik′
day (before yesterday)	bo-ri′	bo-kri′	pong-wo′	pon͡g-wosh′-ko	ki-bi+buk′
dead	dŭ-wa′-wa	———	ke-tong′	———	košht′-ka
deep (water)	di-kru+tyng′-ru	di-kru+tyn͡g′-bru	i-rong′		
deer (large)	mno-dŭ-bi′	———	prun′-mi	shu-ring′	su-tu-rik′

* In the same wide sense as in Bri-bri.

deer (small)	do-ĕ′	——	shu-ring′	shu-ring′+ma	bušh
to depart				to′-to	
devil	bi	——	au	auh	kag-bru′
dew	mo′+ri-u	——	tom-bor′-ia		
to die			ba-ke′-to		
different	hau′-ri	——	o′-bri	ba-bok′-so	bu-i-shi′
to dig	i-biu′-ga	——	koz′+ung		
direction	wo′-mo	——	jou′-re	yow′-re	
dirt	i′-shku	——	wri-wri′-a		
to dive	b-tsat-ku′	——			
doctor	ha-wa′	——	i-bi′	——	
dog				shi′-ti	aušh
to drink	i-ju′	——	i-oz′-ong	——	i-jang′
drop	i-wo′+riu	——	si-ra′		
drum	sŭ-bak′	——	zhung	——	
dry	si′-na	——	i-dog′-i-do	i-dog-roi′	tsen′-ka
dust					ta-brung
eagle	sar′+po	——	do+pung′	——	
ear	zgo-ku′	kwush′-ki	kwong′-wo	kwong′-wa	kwa′-ga
early				ko-o-bi′-ta	
earth (soil)	hi′-zhuk	hi′-zhku	krung	krung′-sho	tap
earthquake	i	——	yo	yo-yan′	
to eat	jĕ-ka′-gri	jĕ-ku′	i-e′	——	ja-ă-de′
eddy	di+ha-ra-mi′	di′+men-e	hi-rung′	pŏ-rung′	
egg	i-si′-a	——	sen′-wa+wa	——	kup
either	mĕ-no′	——			
elbow	u-ra+o′-gu-chi	——			chušh-che′-wi

English.	Cabecar of Estrella River.	Cabecar of Coen River.	Tiribi.	Terraba.	Brun-ka.
empty	wu'ji-ka	———	i-ro-ri'	i-ro-re'	
end	i-bĕ-ta'	———	kok+shur'-bo		
ended				fir'-kru	
enemy	sĕ-na-ma'-ru	bor-gi'	mor-i-mos'-kwo		
enough	wir'-u	er-u'	kwont'-soi	i-ont'-so	
evening	tsa'-na	tsan'-li	kok+sing'-ĕ	———	
every	ben'-a	se-hen'	pir'-gru		
eye	wo'-bra	———	bok'-wa	———	kai'-ish
face	wo'	———	bok-woi'	bok-wo'	kai'-ish
to fall	i-han'-a	i-han-a-te'	ro-noi'	bo-go-noi'	tru-ŭ-lin'
family	jai'-wa	———	pei'-ga	wai-you'-sa	
far	ka-mi'-mi	———	krong'-e	fo-lai'	ka-wi-gri'
father	i-ji'	ji	kok	———	je-bet
father-in-law	dŭ-bai'	———	ke'-gi	———	
fat (grease)	or'-i	ur	kio'-yo	———	i-bu'
" (animal)				sho-shoi'	
fear				ko-ban-kret'	
feather	i'-ku'	———	kor'-ga	———	ka
to feel	pa-su'	———			
female					i-ra-matk'
fever	kro-na-bi-te'	———	b-gun'	swo-ro'	che
few					ble-a'-dri
to fight	nyak+pu	———	ing'-kwi	i-no-kwe'	
finger	u-ra'+sku	———	sak-wo'	sak'-wo	ku'-skwa
fire	jĭ-ko'	———	yuk	———	ji'-kra
fire-fly			skon'-kwa	skon-kwa'	ki-us'

first	suk'-u-tu	———	bom'-gung		
fish	mi-ma'	———	ma'	———	ung
fish scale	ni-ma'+kwo	———	ma+zre'-kwa	ma+kwo'-ta	
flat	shpat'-ke	———	beg'-ĕ	———	
flea	d-ki'	———	kiung	———	chik
floor	har'-ka	hu+ha'	u+pkwe'-ru	u-i'-ro	
flower	kar+wus'-kwa	———	kor+i-a'-gru	kol'-ong	krang+sret
fly	shi'-bu	———	ji-wir'-wa	———	srung
fog	mo	———	pong	———	
to follow			ir'-gu		
food	jĕ-ku'	———	dli	———	
foot	kru+kwe'	———	shkor'-mo	shkon'-mo	kras'+kwa
forehead					wa-tušh'
foreigner	jĕ-ku'-sŭ-wa	———	po-las'-o	fo-las'-o	
forest	kar'+ga	———	kor'+go-rui	krosh'-ko	
free			tok'-sai	———	
friend	her'-bui	———	pei	pfei	
frog	bur'-bi	———	ŏ-rung'		
tree-frog	d-cha'-bu-kwi	bu-kwi'	knos	kro-wem'	
from			i-res'		
full	tŭ-ri'	———	tu'-nai	ta-bo-tu'-nai	kut-ka'-li
to gather	nya'-ska	———	sha-ras'-o		
ghost					i-wik'
girl	{ bu-si'*	a-la-bu-si'	wa+wa-re'	———	i-ra-matk'
	{ jo-ba'*	———			
to give	i-mak'	———	twoz'+ung		
glad			wo-ko'bi		

* Same distinction of age as in Bri-bri

English.	Cabecar of Estrella River.	Cabecar of Ooen River.	Tiribi.	Terraba.	Brun-ka.
to gnaw			po-we′-a		
to go	shku	——	pan-o′-ma		
God	si-bu′	——	zi-bo′	zŭ-bo′	si′-bŭh
good	boe′	boi′	kob′-e	{ kob′-soi kob′-ĕ	} mo-reng′-ri
grandfather	ta′-chi	——	bog-un′	——	
grandmother	ho-jĕ-ku′	——	ter	——	
grass	ta′-tsik	——	jit-sho	kro′-sho	
green	spa	spa-na	ke-song′	ke-song′-ĕ	
to grind			zguiz′+ung	a-we′	
to grow			kun	——	
guatuso				shku-re′	chešht
gun	mok′-kur	——	krik	——	bot
hair	to-kung-gu′	——	kog′-zn	kong′-zn	sišh′-ka
" of body				zo	shi
half	mos′-ka	——	wen-zon′	wor-bish′-ko	išh-ta′
to hallo	m-han-o	——			
hammock	ki-pu′	——	po′-gro		kung
hand	u-ra′	——	ork′-wo	——	ju-re′
back of hand	u-ra′+shi-bi	——	ork-wo+king′		
palm of hand	u-ra+ktu′	——	ork-wo+kwo′-broi		
handle	i′-ka-ta	——	kroug	ba-sor′-ko	
to hang	i-kar′-mi	——	priz′-kom-o		
hard	d-re′-re	——	koug′-ko	kang′-kwe	
to have	krir	——			
hawk	pung	——	u-re′+pung	pung	suñg

he	hé	———	we	kwe	i+a-bi′
head	dzĕ-kung	———	kog′-o	———	sa-gra′
head of tree	kar′+u-ra	———	kor′+de-woi-rung		
to hear	his′-ka	———	kuz′+ung	ko-kuz′-ung	
heart					kwi-sišh
heat	ba	bar	bol-o′	kli-kli′-a	ji+ong′
to help			kim-toz′+ung		
here	is′-ka	———	husk′-ko	ir-ish′-ko	we-eh′
high	hu-i′-ne	kong+shu′-ki	kok′+shko	kok+shko′-i	ašht-kar-e′
hill	kong+tsu′	———	drub	———	kak-tušht′
to hold	i-kru′-ku	i-kru′	shaz′+ung		
hollow	bar	———			
honey	bur	———	or	or′+di-o	but+cha
horn	i-du-ra′	———	su		
house	hu	———	u	———	uh
how	m-ne′-kĕ-pi	———	so′-je-ra	zhi′-ri-a	
husband	(hĕ+)ji′-ji	———	(bor+)do-met′	(bor+)do-ben′	
hunger	bĕ-chi′-na-te	ka′-gro	pli′-do		
I	jis	———	ta	———	a+da-bi′
idle	bi-ki	———	pag-je′-me		
iguana	bo′-a	ba	wong	fwang	ă-lit′
inside	hi′-na	———	ir′-or-e	ir-osh′-ko	
intestine	nya	———	jang′-ko-gru	zhang	su-a′
iron	ta-be′-ri+ji-ka	ta-be′+ji-ka			
jar	hung	ung	zbi	———	
jaw	kab-sha′	———	o-ro′	———	
jigger	d-ki′+tsin-la	———	kiung′-wo	———	
juice	i-diu′	———	di-o′	———	cha

English.	Cabecar of Estrella River.	Cabecar of Coen River.	Tiribi.	Terraba	Brun-ka.
just			so-kob′-so		
knee	k-chu′-wu	———	ko-kwo′	ku-wă′	
knife	ta-be′-ri+je-ba	———	su-gro′	druñg	man-kra′
knot	i-wo′-moa	———	ki-bosh′-kwe	———	
to know	hu-nyer′	che-mi′	mi-dep	mi-der′-a	
language	suk-tu′	———			
large	kŭ-bi+bri-wi′	———	tu-e′	kwon-e′	
very large			kiz-bong′-ĕ		
last	shin′-a-wa	———	jam-o′		ok-sha-re′
late	ka-mas′-kra	tsan′-li	kok+sing′-e	———	
to laugh	m-han-ya′	———	har		
lazy	bi-ki	———	pag-ji-mo	kar-a-guñg′-e	king′-shu
leaf (of tree)	kar+gu′	———	kar+kor′-gu	kor-o′-ga	krang′+ka
" (broad)	sik	———		zor′-go	
lean	j-kai′-bŭ				
to leave			ber-i-oz+ung		
left hand	shu-rĕ′	———	mir′-e	———	
leg	kru+kra-be	———		kwos-bo	
lemon	ash′-ua	———	ĭ-gen-moh′		
to lend	pei′-te	———	poz-kwoz′+ung	———	
level	shpat′-ke	———			
to lie	b-kan-ju′-wa	———	nyor′-ke	———	ka-tu-a-ri
to lie down	m-he-bĕ-kru-wa	———	tush′-ko	bu+tush′-ko	ja-ung
lid	i-kush′-tu-gru	———	bong-za′-kru		
to lift	ĭ-bĕ-tsu′-ku	———	po-yo′-zu	po-yoñg′-zo	
lightning	kong+wo+hor′-ku	———	zhgu-ring′	zhu-ring′	ji′-kra

lips	ko′-kwu	—	kap′-kwo	—	
to listen	i-ken-su′	—	kuz′+ong	ko-kuz′-ung	
little	tsi′-nĕ-kra	—	ti′-ra	so-ti′-ra	i-sta-mu′-ra
liver	her	—	wŏ	wo	kok
long	kar-kwe′	b-tsi′		bu	
to look	sŭ-wa′	i-sung′	i-zung′		wa-izh′-da
to loose	i-kuk′+tu-wa	—	kŏngz′+ung		
to lose	i-kit′-sen-a	—	hen-oi′	—	a-ring′-ra
louse	kung	—	—	—	
lump	wo-wut-ser-e′	—	zo-i-ring′	kwo-i-ring′-soi	
lungs	jo	—			
macaw (green)	kwa+si-an′-e	pa	yĕ-mos′+kwong	—	
“ (red)	kwa+mat′-ka	ku-kwa′	kus′+kwong	kish′-kwong	
maggot	d-chun′-ya		shto′	—	
to make	i-jo-wŭ′		she-ri-oz′+ung	sho-ri-oz′-ung	
male					kong+ad′ra
man	hĕ-ji-ji′	wi′-pa	dŭ-met′	do-ben′	kong+roh′
many	o-ru-i′	—	et′-o	yon′-soi	et-sušh-ta′-ri
how many?					bi-ik′
mark			i-ya′		
marsh	do-ri′	—	krung′-sho	—	
meat	jĕ-ku′+gru	jĕ-ke′	sing	dŭ-li	i-u-ran
medicine	buk′-pu-ri	k-pu′-li	du-ro′	—	
to mend	i-wo-yŭ′	—	yĕ-woz′+ung		
middle	shas′-ka	shus′-ka	wor-be′	wor-be′-shko	išh′-ta
midnight	nĕ-nye′	kong+shong-shong′	kog-rong′-ĕ	ko-i-rong′-i	kag+išh-ta
milk	tsu′+diu	—	nau′+ri-o	—	tur-i-ka′

English.	Cabecar of Estrella River.	Cabecar of Coen River.	Tiribi.	Terraba.	Brun-ka.
mine				to-nya′	a-rek′
monkey (red)	sar+mat′-ka	——	du′-i+go	do	non͡g
" (howler)	d-ke′	——	bib′+go	bib	u′-li
" (white face)	kuk	——	yai′+go	——	ok
month				mok+ra-ra′	
moon	to-ru′	tu-ru′	mŏk	mok	tĕ-be′
more	i-ki′	——	o′-bi	——	
morning	si-nyi′-ra	——	kok+shrung′-to	kok+shrung′	
mosquito	di-kri′	——	kwis′-king	kwis′-kwing	srun-sit
mother	mi	——	meh	——	ba′-bi
mother-in-law	jak	——	aim	——	
mouse					kwi
mouth	ko′-kwu	ku′-a	kam′-o	——	kas′-a
mouth of river				di+sor′-go	
much	bi-ku′	——	et′-o	yon′-so	
mud	do′-ri	do′+ji-ka			so′-ra
nail (finger)	u-ra+sku+sho′-kwe	——	sak-wo+drung′-yo	——	drig′-i
narrow	si-ma′-na	——	pro-tir′-a	——	ku-ha′-dri
navel	mo′+wo	——	tu′-wa	tu′-won͡g	
near	ko-ku′-na	——	sosh′-ku	——	ek-chish′-ĕ
neck	ku-ri′+ki-cha	——	kring′-dok	——	ing-sa′
necklace	ku-ri′	——	bra		
night	nĕ-nye′	——	shke	——	big-at′
nipple					ka
no	kai	——	je′-mĕ	zhe′-mi	
noise	ko′-ju-gĕ	——	tu-:-et′-o	i-rong-ke′	kshis′-ka

noon	mo′-ki	———	dlo-drub		kag+i-bušh′-ta
nose	jik	———	ne′-kwo	———	kshis′-ka
not	kai	———	zham′-ko	zhe′-mi	
now	hir	———	er′-i	———	cha
numerals					
1	et′-ku	———	kra-ra′	kra′-ra	et-sik
2	bot′-ku	———	pug′-da	kra′-bu	bug
3	m-nyar′	m-nyat′-ku	mÿa′-ro	kra-mi-a′	mang
4	kier	kil	p-kegn′-de	kra-bŭ-king′	bašh′-kan
5	sker′a	———	shkegn′-de	kra-shking′	kšhi-skan′
6	ter′-lu	———	ter′-de	kra-ter′	tesh′-an
7	kul	———	kog′-ŭ-de	kra-kok′	kušhk
8	pagl	———	kwog′-ŭ-de	kra-kwong′	ošh-tan
9	te-ner′-lu	———	shkow′-ŭ-de	kra-shkap′	
10	d-bom′	———	dwow′-ŭ-de	kra-ra-wab′	
11	d-bom+et′+ku	———	king-shu+kra′	king-sho+kra′-ra	
12	d-bom+bot′+ku	———	king-shu+pok′	king-sho+kra-bu′	
13			king-shu+mya′	king-sho+kra-mi-a′	
14			king-shu+p-kegn′	king-sho+kra-bŭ-king′	
20			dwow-ŭ+pug′-da	sag+puk′	
30				sag+mi-a′	
oil	ki-u′	———	ki-o′	———	
old (thing)	i-ken′-a-wa	———	ke-ge′	zan-fe′	so-gro′
" (person)	i-ken′-a-wa	———	yi-no+ke-ge′	zan-fe′	
on			hong	č-li′	
once				fra-ra′-soi	
other					e′-je

English.	Cabecar of Estrella River.	Cabecar of Coen River.	Tiribi.	Terraba.	Brun-ka.
otter	di′-hu-a	———	won-yuk′	wan-yŭ′	
our			shi′-ya		
out	i-we′-ten-ŭ-wa	shi-bi′-ga	kok′+su	hu′-ga	
outside	wi′-tu-ki	———	kok′+sur-e	hu′-ga	
over	king′-ga	———	kiu′-go		
to own			toi		
package	dŭ-li′	———	shwong′+i-do	d-bok′-tan	ets-ong′
pain	di-ka′	di-li′-na	du-re′	ban	sa-o-ra
peccary (wary)	sir-bi′	———	shi-ri′	———	kra-mi-shuk
" (small sp.)	kas′-ri	kas-i-ri′	shtuk′-o	———	si-ui′
piece	i-wu-ku′	i-wu′-gi-tu	ba-ja-mo′	bo-kwa-ra′	
pile	i-dŭ-pan′-a	———	drub-par′-a		
to pinch	i-bĕ-tsu′	———	shoz-goz′+ung	———	
pine-apple	ka-ru-ru′-bui	ha-mu′-wa	pong′-wa	———	bwat
plantain	kru-bi′	———	i-bing′	———	mwa
point	i-bĕ-ta′	———	d-woi-rùng′	d-bor-ong′	i-dru′
pot	ta-be′+wi	ta-be′+ung	droz′-bi		kwa-le′
precipice	hak+en′-a	———	i-geng′	i-geng′ skin	
priest	ksu′-gru	———	tang′ra		
to pull	si-ku′	si-gu′	shiz′+ung	———	
to push	i-nya′-tu-wa	pat-ku′-ju-mi	poz-rez′+ung	———	
quarter	i-ju-wa′	———	ba-bog′-yu	bob-wo-king′-de	
quick	ho-ru′-si	———	mal-ish′-tĕ	mal-e′	do-do′
quiet	kai-wa-shu′-ya	———	king′-e		
rain	kan+i′	———	shu-nyo′	shu-nyong′-wa	jo

rainbow	shko-ba′	————	dŭn		
raw					tašh-ka
ready	bi-ka′	b-tsi′	ta-pid′-zung		
red	mat′-ka	————	sri-zrin′	kro-sri-zri-ren′-o	kru-bat′
reddish	b-tse′-na	————		sri-ren	
region	sar′-as	kin	kosh′-ko	————	
to rest	he-ru′	————	wosh′-ti	wosh′-to	
to return	re′-mi	bra′-ni		zwor′-kwoz-ung	
rib				zhgring	
ribs	pre′	————	sheng′-ro	zhgring′-ro (special plur.)	
right hand	mo′-ki	————	kil-e′		
ripe	uri′-ru	————	kwi	————	brot-ka′-li
to rise	ku′-kn	————	shwo′-zu*	————*	
river	di-kru+e-na	di-kru+tyng′	di-kes′	————	di+kak′
road	nya′-ra	————	ir′-o-bo	————	neng͡-kra
roof	hu′+tsang	————	ur+bo	————	
roots of tree	kar′+gi-cha	————	kor′+su-wo	kor+sreng′	kshas
rope	ksa	————	kyung͡		
rough	pa-nen′-e-ir-u	pa-nen′-e	ŏ-nos′-teng		
round	bri′-na		ki-ring′-ĕ		phi-gi-dat′
to run	b-ton-o′	bi-ton′-u-mi	ti-ti-čh′	————	
sad				woi′-di	
saliva	kus′-kwu	————	trung	————	
salt	dĕ-ji′	————	drung	————	ki
sand	ksong͡	————	čra-sho′	————	up
sap	i-di-kli′	kar′+diu	kor′+di-o	————	
to say	i-shu′	————	o-ros′	tre	

* Imperative singular.

English.	Cabecar of Estrella River.	Cabecar of Coen River.	Tiribi.	Terraba.	Brun-ka.
fish-scale	ni-ma′+kwo	————	ma+zre′-kwa	ma+kwo′-ta	
to scatter	sho-ti-a-wa-mi	tu-ru-mi′	zhduz′+ung		
scorpion	bi-che′	————	di-ye′	————	
to scratch	hen-biu′	biu	zhguz′+ung	————	
sea	dĕ-je′	————	drung	drung′+shko	ki
to see	sŭ-wa′	————			
seed	kar+wu′	————	kor′+kwo-wo	————	
to send			i′-ri-os	i-ti e′-ba	
to sew				sez′-ung	
shadow	wig′-ra	————	ya′-gro	sen′·te	ka-wik′
to shake	i-wuk-bu′	————	yoz′-ung	————	
shallow	pre′-na	pre′-rĕ	ir-o-ti′-ra	————	
sharp	i-wu-ku′	a-ka′+ta	su-ret-eng′	{ (edge) peg-you′-soi (point) su-ret-eng′-e	
she	he	————	we	kwe	
shell (bivalve)	su-ri′	————	{ wu-rir′-kwo shuug	(murex) flin′-kwa dŭ-ram′-kwa	
shield	so′-gru	————	sob′-kwo		
shin	kra-be′	————	shkon′-gro		
to-shoot	i-tiu′	————	io-roz′-ong	————	
short			kwo-tir′-a	————	
shoulder	ko′-ka	————	d-wob′-dok		
shrimp	sok	————	kuz′-gwo	————	
sick	kri-na′-wa	————	sur-o-eh′	sur-o′	che-at′
similar	he′-kĕ-pi	————	en′-i-yŭ	sog-bak-teng′-e	i-dosh-ĕ-re′
to sing	bk-su′	————	toz′-ung	————	

to sink	i-wo-tun-a′	——	piz′+ung		
sister	ko-ta′	——	dor	——	
sister-in-law	shdo-bu′	——	kak	——	
skin	kwo	——	kwo′-ta	——	kwas-kwa′
woman's skirt	ba′-na	——	shwom′+zring		
skull	tsu-ko+du-chi′	dze-kung+du-chi′	ka′-dre-a	kog-wo′+dŭ-bo	
sky	du-ra′-bru	——	kob′-kwo	kom-ong′	
to sleep	sku-ri′-na	i-skwu′	poz′-ung	——	kap
sloth (brown)	si-na′-si	——	sho-rong′	——	
" (gray)	se′-ri-a	——	ki-a′	——	
" (little)	di-klu′-ji-ba	——	suz-bo′	——	
slow	hen′-a	——	wor′-a	——	ŭ-wa-ta′
small	tsi′-nĕ-kra	——	tĭ′-ra	so-tir′-a-wa	i-sta-mu′-ra
to smell	ha-rar′-ge	——	i-ro′-shis		
smoke	shkon-ĕ′	shkon-o′	nyo′	——	ji-ĭ-ja′
smooth	jes′-na	je	tu-it′		
snail	*bu-ri	——	*pu-le′+na-kwa	shku′-ra-kwo	
" shell	*jok′-sĕ-ro	——	*i-wi′-bru	du′-kwa	
snake	kĕ-bi′	——	b-gur′	——	ta-bek′
to sneeze	tser′-gĕ	——	shor-ke′		
soft	do-ru′-na	——			
sole of foot	kru+ptu′	——	shkon′-kwo	ku-kwo′-woi-ro	kras-kwa+plang͡
solid			kong′-e		
son	jĕ-ba′	——	wa	——	
sour	shkun′-a	——	shpo-et′-o	shpo-you′-soi	
spider	si-a′-mu-nya	——	ting		
to spit	kar-i′	——	trung+woz′+ung	trung+twoz′-ung	

* Probably special names of *Helix*, *Bulimus*, *Melania*, &c., as in Bri-bri dialect, *kwa*, in Tiribi evidently means shell.

English.	Cabecar of Estrella River.	Cabecar of Coen River.	Tiribi.	Terraba.	Brun-ka.
spleen	ko-u′	tak	woug	———	num
to spread			kwob-toz′+ung		
spy	ju-jet′-kuk	———	prob′-a		
to stand	her-wu′	———	jon′-ke	zhong-zhong′-ke	
star	koñg+wu′	———	d-bar-bo′	d-be-ra′-kwo	um′-ra
to steal	hog′-bru	———	ru-kez′+ung	rur′-ke	a-jañg′-li
stick	kar	———	kor′+lo	———	krang
to sting	m-her′	ka′-wa	ta-wa′-ru	shor-ke′-zo	
to stink	sa-ru-i+ha-rar′	su-ru-i+hrar′-ge	o-no′-et-o	o-no-oi′	
stone	hak	———	ak	———	kang
stool	krŭ-wa′	———	kruk	kru	
stop	č-re′-re	he-re-re′	ko-shoz′+ŭ	———	
straight	mog-i′	———	i′-teng	kro-bek′-oi	du-u′-ka
to strike	it-pu′	———	i-ruz′+ong	sphoz′-ung	
strong	d-re′-re	———	d-bo-e′-to	kang′-ku-i	s-du-ru-re′
sugar	pash-tu′+chi-ka	———	jir′-bo-sho	sror′-bo-sho	
sun	di	di′-wo	do-ro′	———	kak
sure			zog′-u-de		
to swallow			d-kwoz′+ung	d-woz′-ung	
to sweat	pa-re′-na	———	sho′-ri-a	sho-ri′-na	
sweet	bro′-na	bro	pre′-bre	———	
to swim	bo-wo-ku′	bo-kwu′	ah-weh′	we	a-bu′-ŭ-li
tail	i-m-nek′	———	proc	frac	kwan-tsa′
to take	i-tsa′-mi	i-tsu′-mi	kroz′+ung	———	
to talk	bk-tu′	———	tnoz′+ung	———	
tall	b-tsi-br-wi	———	kok+shko′	———	

tame			un-e′-bi	———	
tapir	na′-i	———	so	———	na-i′
to taste	tso͡ng-sir′	———	kuz′+ong		
teeth	ka	———	ko-wo′	———	kas′-a
" (molars)		ka+nyuk	ap	———	kam′i
to tell	i-shu′	———	ir-o-tos′		
then	he-re′-re	he′-re-re	en′-i	er′-a-ga	
there	wa-ji′-ga	di-ya′	kem′-nŏ	kim′-rish-ko	
they	he′-wa	ye	kweb′-ga	eb′-ga	i-rošhk
thick	kro-na-bi-ti′	hu-re′			
thief	hak	hak-bru′-gi			
thigh	tu-bru	———	kwor′-wo		
to think	her-wik′	———	wo-tuiz′+ung	woi′-du	
thirst	her-ba′-na	her-si′	baum′-gung		
thorn					kwa
thou	ba	———	pa	fa	ba+a-bi′
throat				beng′-so	
to throw	i-hu-wa-mi′	i-tu′	d-buz′+ung	———	
thumb	u-ra+sku+in′-ga	———	sak-wo+kes′-yu	sak-wo+kes′-i	
thunder	hŭ-ra′	a′-ra	krik	———	kak′+an-te-gra
tick	ban′-o-wa	kas′-o-wa	tri′-gua		
to tickle	gi-pa′	———	yi-gru-woz′+ung		
to tie	i-mo′-a	———	kro-di-oz′-ung	kro-tenz′-ung	
tiger (spotted)	na-ma+kro′-na	———	d-kro+tong	d-bo͡ng+tan-tan′-e	ku-ra′
" (black)	na-ma+do-ro′-na	———	kro+zi′-a	d-bo͡ng+kro-si′-a	
" (red)	na-ma+mat′-ka	———	kro+sri-zrin′	shu-ring+d-bo͡ng′	
time (future)	ma-nya-ui′	nya-nya+si-nye′-ta	pĕs-ir′-o	de-nash′-ko	

English.	Cabecar of Estrella River.	Cabecar of Coen River.	Tiribi.	Terraba.	Brun-ka.
time (past)	ba-ha′-si	nya-nya′	den-ai′	po-nyash′-ko	
tired	hě-rě-be′-na	———	sen-oi′	———	j-ro′-sha-ra
toad			krok	kro′-we	kshuk
tobacco	dŭ-wa′	———	do-wo′	———	du-a′
to-day	hir	———	er′-i	———	cha
toes	kras′-ku	———	sap′-kwo	———	
together	mya′-ra	———	hu′-nĕ	sha-ra′	
to-morrow	bu-ri′-ri	———	i-bong′	i-bou′-a-ga	sek
tongue	kok′-tu	kuk′-tu	kior′-kwo	ker′-kwo	kwat′-kwa
top of head	tsŭ-kung′+bě-ta	tsong′+bě-ta	kwa-dru-i-bo′		
tortoise	kwi	———	kwe	bog-o′	ba-eg′-ri
to touch	bŭ-rat-ku′	———	por-woz′+ung		
tree	kar	———	kor	———	krang
tribe	wag′+ĕ-ru	———	io-no′	———	
trunk of tree	kar	———	kor+i′-do		
to try	he-man′-i	———	tŭ-nit′		
to turn	men-e′-wa	———	wor-kwo′+zu		
twice					bug-ŭ-dek′
to twist			doz′+ung	zhu-iz′-ung	
ugly			so-o′-wa	———	
ulcer	shum-e′	———	zbing		
uncle	no-wa	naung	ke′-ga		
under	d-king′-ga	———	i-dor′-ko	tush′-ko	a-de′-bi
to understand	jo-na-he′-ra	te-sa′	ku′-rob-dě	ko-kuz′-ong	
unripe			kro-pro-pru′-gi		
until			pe-si′-re		

valley	ko͡ng+d-choi′	———	ztomb	zdam	
vein	ham-e′	———			
vine	ksa	———	kij′-wo	———	san-kwa′
voice	har′-ke	———	krun		
to vomit	i-tu′-wa	———	yar-we′	ya′-twe	
waist	ki-pa′	———			
to wait	kin-su′	———	ko-shoz′-ung	———	
to walk	i-shku′	———	hik	ik	
to want	ki-a-na′	———	woi-det′	———	
wasp	bu-kra′	———	wŏ-ro′	wo-rok′	srung-wa′
water	di-kru′	———	di	———	di
wax	har	bur′+i-nya	sot	on′-cham	but
we	sĕ-ha′	———		shing	ja-a-bi′
well	boe′	———			
wet	kis′-u	nun-a′-ga	puk-tong′-ĕ	pu-hu′-i	se-kar′
what	hi-ru′-in	er-hi′-i	kon-e-kro′-ti	zhe	
when	mi′-ga	———	jon′-wro	zhon′-wa	i′-ik
where					je
which	men-ĕ′	———	kon-e-de′		
to whisper	ka-kwa′	jĭ-kwa′	ba-tnez-wor′-a	kob′-wo+ba-kwoz-ung	kos′-wa
to whistle	ka-kun′-e	ka-kun′	b-ko′-ba-kwe		tsišh-kung
white	su-ru′-ru	———	plu-blun′	kru-ru′-ni	su-wat′
who	ji	———	i	e	di′-a
whole	har-e′-bu	kra-be′	kwo-hik′	i-don′-hi	
why					io′-ge
wide	shon′-a	shon	ba-met′-o	star′-e	wan-ka-li′
wife	(je+)ĕ-ra′-kra-wa	(je+)ĕ-ra′-kra	(bor+)iok′	(bor+)wa-re′	i-ra-rok′

English.	Cabcear of Estrella River.	Cabcear of Coen River.	Tiribi.	Terraba.	Brun-ka.
wild	kar+ŭ-bu′	———	kro-hi′		
wind	si-wang	———	pruc	———	shung
woman	ĕ-ra′-kra-wa	ĕ-ra -kra	wa-re′	———	i-ra-matk′
worn	d-cho′-nya	———	chus′-kro	———	
worse	se-ru-i+si′	———			
to wring		diu′-tru	zhguiz′+ung		
wrist	wu′-gĕ-cha	———	ork-wo+dok′		
yellow	psi	———	shoin′-lot	shoi-rang͡g′-i	sho-o-sat′
yes	hŭ′-ŭ	———	ing	———	u′-ge
yesterday	jĕ-ki	———	kub′-kĕ	kub-kesh′-ko	bi-ik
you	bas	be	ta-wa	fai-rung′-kwo	bi-rošhk
your			poi	to-nya′	

www.ingramcontent.com/pod-product-compliance
Lightning Source LLC
Chambersburg PA
CBHW031440280326
41927CB00038B/1289

* 9 7 8 3 7 4 3 3 9 5 3 8 1 *